Imagining Imaging

Imagining Imaging

Michael R. Jackson

CRC Press
Taylor & Francis Group
Boca Raton London New York

CRC Press is an imprint of the
Taylor & Francis Group, an **informa** business

First edition published 2022
by CRC Press
6000 Broken Sound Parkway NW, Suite 300, Boca Raton, FL 33487-2742

and by CRC Press
2 Park Square, Milton Park, Abingdon, Oxon, OX14 4RN

Library of Congress Cataloging-in-Publication Data

Names: Jackson, Michael R., author.
Title: Imagining imaging / Michael R. Jackson.
Description: First edition. | Boca Raton : CRC Press, 2022. | Includes bibliographical references and index.
Identifiers: LCCN 2021025978 | ISBN 9780367428549 (hardback) | ISBN 9780367427825 (paperback) | ISBN 9780367855567 (ebook) Subjects: LCSH: Diagnostic imaging. | X-rays.
Classification: LCC RC78.7.D53 J32 2022 | DDC 616.07/54--dc23
LC record available at https://lccn.loc.gov/2021025978

ISBN: 9780367428549 (hbk)
ISBN: 9780367427825 (pbk)
ISBN: 9780367855567 (ebk)

DOI: 10.1201/9780367855567

Typeset in Times
by Deanta Global Publishing Services, Chennai, India

Contents

Foreword

I am a believer of the hidden parallels between art and science. After all, most of my career has been defined by crossing the boundaries between the apparently independent disciplines.

I've had the pleasure of knowing Dr Michael Jackson since we gave a lecture together at the Edinburgh Science Festival 2017, when he first presented his exploration of the relationship between visual arts and medical imaging, the key themes of which are expanded upon here.

In this book Michael probes these boundaries, looking to rediscover and substantiate common threads and bringing together a comprehensive overview from the early beginnings of art to present-day imaging. Examining a diverse range of artistic traditions and movements, all manner of popular culture references, and combined with insights from his own expertise as a clinical radiologist, we find links between medical imaging and creative visual representation in a wide range of unexpected arenas. Previously compartmentalised art and science topics acquire new relationships during Michael's bridging analysis, observations, and debate.

Imaging techniques compel us to engage with the fragility and wonder of the world around us on all levels. Can we redefine the role of art and science as the same pursuit of truth, when expressed by the artist and explained by the scientist?

Hugh Turvey, HonFRPS

Acknowledgements

This book would not have been possible without the support and resources provided by numerous organisations and individuals. I am grateful to CRC Press for publishing this rather idiosyncratic work by a first-time author. Being neither historian, art historian, artist, nor art critic I am arguably ill-qualified to have embarked on this undertaking, but spending my working life looking at images gives me a stake in this field and I am thankful to CRC Press for recognising this. Kirsten Barr has been both patient and flexible whilst ensuring I (largely) met my deadlines.

Although not a historian, I was fortunate to have undertaken a year's study at the Wellcome Historical Unit as a medical student which nurtured an interest in how health and disease have been framed over time – aspects of which are explored in a near literal sense within this book. William Bynum, my supervisor for that year, and his wife Helen provided impetus to this project early stage. The Wellcome Collection also provided a great number of the illustrations.

In more recent years my involvement in the British Society for the History of Radiology has helped me delve into the story of imaging and I am grateful to council members Arpan Banerjee, Liz Beckmann, Brian O'Riordan, and Adrian Thomas for their assistance with this project.

Adam Zeman at the University of Exeter was very helpful in relation to my aphantasia-related enquiries. Fiona Wotton, CEO of Creative Kernow, provided valuable assistance tracking down several elusive papers. My brother, Andrew Jackson, delivered some crucial IT assistance.

The Artiscience Library, based at Summerhall, Edinburgh is a fabulous resource and I am very grateful to Colin Sanderson for his enthusiastic support and providing access to the library. Lockdown restrictions prevented me from visiting as much as I would have liked, but I will be back and encourage readers to do likewise.

I became aware of the Artiscience Library after giving a talk on "The Art of Medical Imaging" at the Edinburgh International Science Festival in 2017. I am also grateful to the organisers of the festival for putting me in touch with X-ray Artist Hugh Turvey with whom I delivered the presentation. Hugh has very kindly provided the foreword, cover artwork, as well as an example of his work in Chapter 7. I am similarly grateful to all the artists who have allowed me to include their work and would strongly recommend readers explore their portfolios online.

My NHS colleagues have been a source of inspiration throughout the pandemic and are a pleasure to work with on a daily basis. In relation to this book specifically, I am grateful to all the radiologists, radiographers, imaging department staff, and clinicians who have contributed images, discussion points and general assistance. Kirsten Kind, my Manchester colleague, provided a critical nudge of encouragement without which the book may not have come about.

This book certainly would not be in existence without the love and support of my late father, Christopher, and my mother, Sheila, to whom my gratitude cannot be adequately expressed.

Finally, I would like to thank my wife Rachael and our sons Nicholas and Daniel whose love and patience were integral to completing the manuscript. They will be pleased to hear I do not intend to use the phrase "mmm … maybe that could go in the book" for some time to come.

Author

Dr Michael R. Jackson is a Consultant Paediatric Radiologist, based at the Royal Hospital for Children and Young People, Edinburgh. He has a longstanding interest in the history of medicine and radiology, and is Trustee of the British Society for the History of Radiology and Archivist to the Scottish Radiological Society. He is the Royal College of Radiologists/British Society for Paediatric Radiology Travelling Professor for 2021-2022. Dr Jackson's public engagements has included two sold-out talks at the Edinburgh International Science Festival in 2017 and 2018.

Dedicated to radiographers, radiology department assistants, nurses, and administrative staff in imaging departments everywhere – the often unsung heroes of medical imaging.

1 Origins

The origins of many medical specialities are largely lost in the mists of time – dating back many hundreds, if not thousands of years. In most respects that is not the case for radiology and the field of medical imaging, which can, according to the most familiar narrative, trace their origin to Friday, November 8, 1895. On the evening of that date (now commemorated on an annual basis as both World Radiography Day and International Day of Radiology) Wilhelm Conrad Röntgen, Professor of Physics at the University of Würzburg, was working in his laboratory and noted unexpected glowing from a screen coated in barium platinocyanide crystals. He was passing electrical current through a coil located within an evacuated glass tube, an established method for generating cathode rays. While cathode rays were known to cause fluorescence in platinocyanide screens, Röntgen had wrapped his tube in black cardboard which is impervious to cathode rays. He immediately realised that the cause of the glowing green light from the paper must be "a new kind of ray".[1]

Keenly aware of the significance of his finding, Röntgen spent the following month meticulously investigating the phenomenon that he labelled X-rays. Through a series of experiments, he demonstrated that they could not only penetrate card but also books, wood, rubber, and a variety of thin metal sheets. Photographic emulsion was fogged or opacified by exposure to X-rays. On December 22, 1895, Röntgen instructed his wife, Anna Bertha, to place her hand on a tabletop and balanced a photographic glass plate on top of her hand. By activating the coil in a Crookes tube located beneath the table (for perhaps around 15 minutes) Röntgen produced an iconic and unique radiographic image of the human body – the bones of Bertha's hand and a metal ring clearly visible, but with skin and flesh rendered translucent (see Figure 1.1).[2]

Röntgen presented a summary of his work to the Würzburg Physico-Medical Society on December 28, 1895, and news of his discovery spread across the globe with remarkable speed, generating excitement and hyperbole amongst the scientific community and public alike. The clinical utility of X-rays was recognised instantly, and the first radiograph performed for medical purposes was taken within a month of the announcement. From that point radiographs (photographic images of the body acquired using X-rays), became rapidly established as a diagnostic tool, initially for broken bones, bony tumours, and gunshot or bullets embedded within soft tissue – structures of high density that could be demonstrated with relatively short exposure times.[3]

Amongst these early radiographs is an example from 1896, taken by the pioneering physician Dawson Turner in Edinburgh. It shows the chest of a three-year-old child who had swallowed a halfpenny, with the coin visible halfway down the chest, stuck within the oesophagus. The exposure time for this image (which shows the coin clearly, the ribs to some extent, but not much else) was 4½ minutes, an

DOI: 10.1201/9780367855567-1

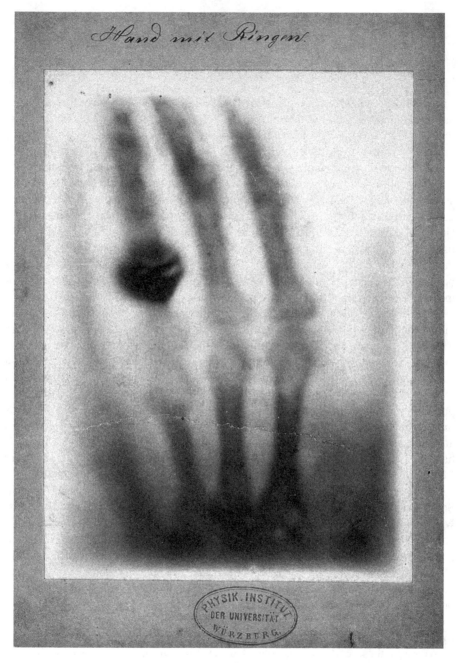

FIGURE 1.1 The bones of a hand with a ring on one finger, viewed through X-ray. Photoprint from radiograph by W.K. Röntgen, 1895. Credit: Wellcome Collection. Attribution-NonCommercial 4.0 International (CC BY-NC 4.0).

impressive feat to have immobilised a child of this age for that long.[4] Fortunately, exposure times have reduced substantially in Edinburgh and elsewhere during the intervening years, even if, as we shall see in Chapters 3 and 8, children's appetite for swallowing coins has not.

Röntgen's work certainly sounded the starting pistol for the emergence of radiography (broadly speaking the endeavour of acquiring X-ray-based images of the body), radiology (the business of interpreting such images to provide a diagnosis or assessment of disease processes) and radiotherapy (treating patients using X-rays and radioactivity), and provides a convenient starting point for our story. However, while Röntgen's work relied on an evacuated tube, it was certainly not conducted in an intellectual vacuum, with several important precedents leading up to November 1895.

In 1785, William Morgan, a British physician, physicist, and statistician, presented a paper to the Royal Society of London on the effects of passing electrical currents through a partially evacuated glass tube, describing the appearance of a glow within the tube. Morgan did not know what was causing the glow, but his report is considered a possible early account of X-ray production, and certainly contributed to the work of Michael Faraday on electromagnetism, and subsequently Sir William Crookes' development of the vacuum tube.[5]

Arthur Willis Goodspeed, professor of physics at the University of Pennsylvania and photographer William N. Jennings unintentionally produced a radiograph of coins and other small objects in February 1890 whilst working on induction coils similar to those Roentgen would use five years later.[6] They did not appreciate the significance of the image at the time, but the cross-discipline collaboration between a physicist working in this field and a photographer is noteworthy.

Nikola Tesla had been experimenting with vacuum tubes since 1887 and is said to have obtained images of the bones of his hand prior to 1895. The Serbian electrical engineer and inventor also documented an association of skin damage in relation to this work in 1892, but did not infer a direct causal link, and did not disseminate his broader findings in this area.[7]

Tesla obtained a number of his vacuum tubes from physicist Ivan Pulyui (variously spelt Puluj, Pului, Puliui, Pul'uj).[8] In his home nation of the Ukraine, Pulyui is venerated as the true discoverer of X-rays,[9] but elsewhere this is contested. Having worked with Röntgen in the same department at Strasbourg University during their early careers, Pulyui would go on to develop his own distinctively shaped "bullet" vacuum tube, which won the silver medal at the International Electrotechnical Exhibition in Paris, 1881.[10] He became a lecturer at the German Polytechnic Institute in Prague, where Roentgen is said to have attended his lectures. He certainly visited Pulyui's laboratory on at least one occasion, when he was presented with one of Pulyui's vacuum tubes.[11]

Ukrainian sources describe Pulyui using these tubes to produce radiographs of a broken arm of a 13-year-old boy and his daughter's hand with a hairpin lying under it, predating Roentgen's radiographs by "a couple"[12] or "a few"[13] years. Unfortunately, the chronology is difficult to establish and these images are not well preserved – much of Pulyui's archive went missing during the Soviet era. Pulyui certainly published a paper in 1889 describing how sealed photographic plates became dark when exposed to something invisible that emanated out of his "bullet" tubes, and earlier work *Luminous Electrical Matter and the Fourth State of Matter* was translated into English by the Physical Society of London the same year.[14] His champions suggest this work describes X-rays as a phenomenon, but the message was lost in an old-fashioned style of writing and in 1901, despite some protestation on Pulyui's behalf, the first-ever Nobel Prize for physics was awarded solely to Röntgen for his discovery of X-rays. In recent years there has been some reappraisal of Pulyui's legacy, and several key insights into the generation and properties of X-rays have been attributed to him including their ionising properties and an understanding of the spatial distribution of X-rays from the site of generation.[15]

There are numerous other figures who can stake a claim in the story of the discovery of X-rays, and it can certainly be argued that the story of radiology has an earlier starting point than November 8, 1895. In this book, I suggest that artistic, creative, and imaginative endeavour have been integral to how radiographic images (and medical images more broadly) are constructed and interpreted. This chapter will go right back to the beginning to provide an overview of some key artistic innovations that are utilised in modern medical imaging, from the earliest human history onwards.

Painting with broad strokes, we will hop back and forth between different continents and historical eras. I hope that by the end of the chapter you will consider the representations of the human body acquired within the scientific discipline of medical imaging to also belong to the tradition of artistic representation. Pietro Belluschi suggests that visual art holds a special place in human civilisation, being the only medium which can be readily understood by individuals separated by hundreds of years or by different languages.[16] When viewed in this context, readily relatable images such as a radiograph of a hand, 3D CT reconstructions of the skeleton, or a midline MRI of the brain all have a natural home within the broad church of representational art.

CAVE ART HAND STENCILS

During the preparation of this text, a number of newly discovered archaeological sites have made a claim to be the earliest known example of human art, and doubtless between writing this paragraph and you reading it some further sites will be discovered, so to make a claim of where and when human art originated is a contentious issue. Cave art depicting human-animal hybrid figures hunting warty pigs and dwarf buffaloes in a limestone cave in Sulawesi, Indonesia, has been dated to be around 44,000 years old. In depicting imaginary combinations of humans and animals this site has been claimed as the first example of figurative art.[17]

However, most sources appear to concur that among the earliest known examples of human art are the depictions of human hands on cave walls,[18] the earliest known example of which (at the time of writing) has been dated as older than 64,000 years. It is a red hand stencil in Maltravieso cave, Cáceres, Spain, and was actually made by a Neanderthal.[19] The cave in Sulawesi also features hand stencils, but these are attributed to homo sapiens sapiens (the subspecies from which we are most directly descended).

FIGURE 1.2   Photograph of prehistoric hand stencils from Cueva de las Manos ("Cave of the Hands") Río Pinturas, in Patagonia, Argentina, dated between 13,000 and 9,500 years. The hands are a mix of dark and light likenesses, representing a combination of hand prints and hand stencils. Image credit: Shutterstock

Figure 1.2 shows a more recent example of prehistoric hand stencils from Cueva de las Manos ("Cave of the Hands") Río Pinturas, in Patagonia, Argentina, dated between 13,000 and 9,500 years old.

Prehistoric hand stencils/prints have been found in all continents and are a near-ubiquitous feature of paleolithic/neolithic art. There are a number of interesting parallels between cave art hand stencils and the first human radiograph. To produce a hand stencil the hand was placed on the rock surface and paint pigment (either manganese-based black, or, more commonly, haematite-based "red ochre") was then blown through a hollow tube (bone or reed) in a diffuse cloud over it, leaving a silhouette image of the hand on the rock.[20] Describing that process in other terms, a projectile agent was discharged from a point source with the anatomical site of interest applied to a planar surface. By selectively blocking the agent, the anatomical structure produced a two-dimensional negative image of the structure of interest. This is directly analogous to the process by which the image of Bertha Röntgen's hand was produced and the description above applies equally to cave art stencils as to radiographs, including those produced using digital equipment to this day.

It is likely that the choice of the hand for the first radiograph was determined by prosaic, pragmatic considerations such as which part of the body could be most readily positioned close to the source of X-rays and held in position for the duration of the exposure. Nevertheless, there is an undeniable and powerful resonance between some of the earliest known human art and the earliest ever human radiograph. Practical considerations also appear to have been at play in the creation of the cave art stencils; left-hand stencils are more common than right-hand images, thought to be because a right-handed individual typically uses their stronger right hand to hold the pigment tube. Of note, it is most likely Bertha's left hand depicted in the first radiograph – the highly visible ring on her fourth finger usually interpreted as her wedding band. Research based on French and Spanish caves also suggests female hands are more commonly found in prints and stencils than males,[21] so Bertha's hand also follows in this tradition.

While I have drawn a technical comparison to highlight the similarity in the method by which hand stencils and radiographs are acquired, I cannot do the same in relation to cave handprints. These were achieved by the same means as baby hand-prints or children's finger painting might be today – direct application of paint (or at least pigmented mud or clay) onto the hand followed by immediate application to the required surface. However, it is worth noting the resultant "negative" and "positive" images of the hand that the two techniques produce – probably the earliest example of image tonal inversion, a process we will return to.

ABORIGINAL ART

Figure 1.3 shows a male human figure painted on a rock shelter at Ubirr, in the western part of Arnhem Land in northern Australia. This Aboriginal art is rendered in what is now commonly called the "X-ray" tradition. It depicts skeletal structures such as the spine, ribs, and long bones of the upper and lower limbs, but cannot be characterised as just a skeleton, with the flesh of the body also represented. This semi-transparent or perhaps cross-sectional scheme is thought to have originated around 2000 BCE, with subjects including sacred images of ancestral humans, and supernatural beings as well as secular works depicting fish and other animals that were important food sources. Later examples include more anatomical detail such as muscle masses, body fat, optic nerves, and features such as the gravid uterus and breast milk in women. Images created after the arrival of Europeans show rifles with bullets revealed inside them.[22]

In the case of a fish, where the shape of the skeleton approximates that of the body in a two-dimensional image, the conceptual leap from directly observed struc-tures to the rendered art is not vast. However, for human and other mammalian figures the process of depicting multiple anatomical structures in this semi-trans-parent fashion, combined with the stylistic flattening of structures becomes both highly sophisticated and highly abstracted. The consideration of how "realistic" the images are becomes irrelevant in the context of an artistic tradition rooted in dream time.

Although clearly predating the advent of radiographic images by a couple of mil-lennia, it is difficult to argue that this "X-ray art" has any direct bearing on medical imaging. However, thematically these works exemplify a number of concepts that will recur throughout this book: the body as a structure of variable transparency; the use of an established schema with its own conventions; the intersection of anatomi-cal reality and human imagination.

FIGURE 1.3 Photograph of a male human figure painted on a rock shelter at Ubirr, in the western part of Arnhem Land in northern Australia. This example of Aboriginal "X-ray" art depicts skeletal structures, such as the spine, ribs, and long bones of the upper and lower limbs, but visible within the soft tissues of the body rather than a skeleton. Image credit: Shutterstock

THE INVENTION OF DRAWING ... PAINTING ... AND RELIEF SCULPTURE?

Another way to describe the method of acquiring a radiograph would be to say that the subject of interest is placed in very close proximity to a planar surface, onto which an image is projected from a point source of electromagnetic radiation. This description could also apply to the quasi-historical account of how drawing and painting originated. The Roman historian Pliny the elder contended:

> We have no certain knowledge as to the commencement of the art of painting, nor does this enquiry fall under our consideration. The Egyptians assert that it was invented among themselves, six thousand years before it passed into Greece; a vain boast, it is very evident. As to the Greeks, some say that it was invented at Sicyon, others at Corinth; but they all agree that it originated in tracing lines round the human shadow.[23]

He goes on to describe the myth of the "Corinthian Maid", Dibutades, and her father Butades, a potter:

> It was through his daughter that he made the discovery; who, being deeply in love with a young man about to depart on a long journey, traced the profile of his face, as thrown upon the wall by the light of the lamp.[24]

Pliny then describes how Butades then filled in the outline with clay and baked it. Athenagoras, the Athenian philosopher and early Christian apologist provides a slightly different account:

> Linear drawing was discovered by Saurias, who traced the outline of the shadow cast by a horse in the sun, and painting by Kraton, who painted on a whitened tablet the shadows of a man and woman. The maiden invented the art of modelling figures in relief. She was in love with a youth, and while he lay asleep she sketched the outline of his shadow on the wall. Delighted with the perfection of the likeness, her father, who was a potter, cut out the shape and filled in the outline with clay; the figure is still preserved at Corinth.[25]

The variation in these accounts is reflected in the numerous ways in which artists have portrayed this story. Both Saurias and Kraton are largely overlooked. A young couple is most commonly depicted, and most frequently it is the woman (usually, but not always identified as Dibutades) who draws around the silhouette of her male lover, but sometimes vice versa. The surface onto which the shadow of the face in profile is projected varies between a rock, an external wall or the interior wall of the pottery, and the source of light is provided by the sun; an oil lamp; a lantern; a blazing torch or a candle. The silhouette is traced using a variety of media including charcoal, chalk, paint, or scored into the surface using a pointed implement. Cupid sometimes makes an appearance to emphasise the romantic element of the story.[26] A fairly typical example of the scene is illustrated in Figure 1.4, an engraving by Guisseppe Bortignoni based on a painting by David Allan – note the ceramic vase to the right of the couple as a nod to the location within the potter's workshop.

While the inconsistencies in the story and its representation are notable, the key elements remain unaltered – the subject of interest is placed in very close proximity to a planar surface, a point source of electromagnetic radiation is used to project an image of the subject onto that surface, and a means of technology is utilised to preserve the image. Again, this is directly analogous to the process of acquiring a radiograph. It is also pertinent to note how Butades transforms the two-dimensional image of the profile into a three-dimensional relief sculpture, an early precursor of photosculpture (discussed in Chapter 5) and ultimately surface-rendered image reconstructions and 3D printing.

The depiction (admittedly in paintings predominantly from the 18th century and beyond) of different means of capturing the silhouette, echoes the confusion of whether it is the origin of drawing or the origin of painting that is being documented. This resonates with the use of both "negative" hand stencils and "positive" handprints in cave art – the message transcends the medium.

FIGURE 1.4   An engraving by Guisseppe Bortignoni based on a painting by David Allan. It depicts a young Corinthian girl drawing the shadow of her lover as cast by an oil-lamp, representing the origin of painting. Credit: Wellcome Collection. Attribution 4.0 International (CC BY 4.0).

RED FIGURE/BLACK FIGURE POTTERY

In relation to tonally inverted images, renowned art historian E.H. Gombrich points out that it is the relationship between two contrasting signals – on/off, light/dark, positive/ negative – that is all important, rather than the choice of which particular signal, tone, or colour used to designate figures and the background.[27] The somewhat arbitrary nature of this tonal choice is illustrated by ancient Greek vases. Prior to around 525 BCE, figurative designs on Greek pottery (vases in particular) were rendered with black figures (painted black glaze) against a red background (the natural colour of the ceramic), a style known as black-figure. This changed when a group of Athenian painters reversed the scheme. The outline of the figures was marked out but then the background outside the margins was painted black, leaving the figures red, but allowing anatomical and decorative detail to be painted over the red figures in black glaze, rather than scoring marks into the clay as had previously been done.[28]

The change from black-figure to red-figure was widely adopted within the Greek sphere of influence as a technical improvement enabling more intricate designs to be achieved. During the transition period "bilingual" vases were produced with a black-figure design on one side and a corresponding inverted red-figure design on the other. A variation on the bilingual vase is shown in Figures 1.5 and 1.6, a kylix

FIGURE 1.5   Photograph of Terracotta kylix: eye-cup (drinking cup) [exterior] ca. 515–510 B.C. Attributed to Pheidippos. The exterior surface of this Greek pottery is rendered in "red-figure" style, depicting a warrior carrying a shield with a bird motif. Image credit: The Met Museum. Public Domain Mark.

FIGURE 1.6   Terracotta kylix: eye-cup (drinking cup) [interior] ca. 515–510 B.C. Attributed to Pheidippos. The interior surface of this Greek pottery is rendered in "black-figure" style, depicting Dionysos, the god of wine. Image credit: The Met Museum. Public Domain Mark.

or eye-cup dated to around 515–510 BCE. Figure 1.5 shows the exterior, showing a warrior holding a shield rendered in red-figure style, while the interior, shown in Figure 1.6 depicts Dionysos, god of wine, using the black-figure method. Although the Attic red-figure technique prevailed, both schemes make visual sense, and if clay was naturally black and a red glaze readily available it is plausible that the black figure scheme could have prevailed.

The resonance between the ancient negative/positive images of hand stencils/ prints and black-figure/red-figure pottery on the one hand and the modern aesthetic of radiographic image inversion on the other might be dismissed as a superficial resemblance, but the parallels of how the technology shapes the image in combination with aesthetic preference are nevertheless striking.

Let's consider the very first radiographic images in a little more detail. Preceding the iconic image of Bertha's hand were the radiographic images Wilhelm Röntgen saw of his own hand when he placed it between his Crookes tube and the barium platinocyanide screen. He saw a shadow of his hand on the screen similar in appearance to the shadow of his wife's hand in the radiograph – that is the bones were very dark, whilst the flesh did not cast much of a shadow. A schematic diagram showing this arrangement from an early radiology textbook is shown in Figure 1.7. This "real-time" imaging of body structures would be developed into the technique of fluoroscopy, and eventually dynamic examinations of the gastrointestinal system via barium swallows and enemas, and of the vascular system via angiograms. For these

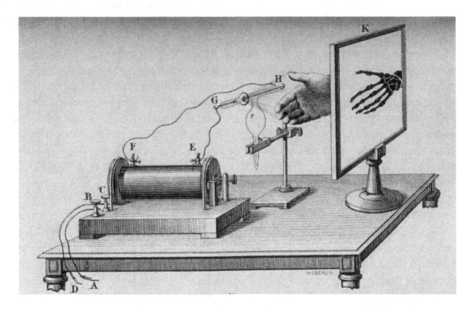

FIGURE 1.7 Diagram from an early radiology textbook (1898) showing the appearances seen on a barium platinocyanide screen when a hand is placed between a Crookes tube and the screen. Note the "bone black" appearance, with the bones of the hand visualised as dark/ non-luminous structures in this setting. Raggi di Röntgen e loro pratiche applicazioni, Tonta, Italo. Date 1898. Credit: Wellcome Collection. Public Domain Mark.

examinations the images (to this day) are displayed in a similar scheme to that which Röntgen saw that first time, with dense structures shown as dark, or "bone black".

This is distinct from the inverted, negative scheme by which radiographs or "X-rays" are most typically displayed. Röntgen's original radiograph was a negative image, but was acquired on a fragile glass plate – he used this negative image ("bone white") to produce positive ("bone black") photo prints to send to colleagues across Europe. Positive photo prints became the conventional format to reproduce radiographs in publications until the 1930s when there was a collective realisation that radiologists reported radiographs in the negative/white bone format.[29] In contemporary digital systems the greyscale can be instantly inverted at the click of a mouse button, but for the majority of radiographic history the image was not rendered in malleable binary coding, but in a physical object, a hard-copy recording of an individual's encounter with the invisible light. It was the physicality of this process which determined the "bone white" for radiographs/"bone black" for fluoroscopy conventions in the same way that the physical properties of the clay and glaze determined the red and black figure conventions of Greek pottery.

This discussion of tonal or greyscale inversion may feel a bit laboured, but there is a fundamental issue at stake. If we can't decide on which image version is the "truest" or "best" in relation to such a stark choice (literally black or white), then we are likely to flounder when we consider more sophisticated imaging techniques in which there are a near limitless range of variables to alter in relation to how an image is both acquired and displayed. Although the negative/"bone white" format has prevailed as the standard means of displaying radiographs for the best part of a century (initially hard-copy radiographs displayed on light boxes, increasingly digital images on computer monitors), the positive/black bone format is still favoured for fluoroscopy or image intensifier-based imaging (often known as "screening"). These reflect historical aspects of how the respective techniques developed – acquiring a static, permanent image as a radiograph involved exposure of photographic emulsion to X-rays, opacifying or darkening those areas exposed. "Screening" relied on those areas exposed to X-rays lighting up. These conventions persist in the digital era, despite the ease with which the greyscale can be inverted. While I would be stretching a point to suggest the bone black/bone white dilemma is a "hot-topic" within the radiological literature, it is interesting to see that there is an ongoing trickle of papers exploring the value of greyscale inversion in a number of different settings.[30, 31, 32, 33, 34]

A hangover of the convention of using the negative ("bone white") display format for radiographs is the potentially confusing use of the term "radiolucent". This is used to describe areas on a radiograph which are less dense than expected, for example, a "radiolucent" area of bone is an area where the bone is less dense and allows more X-rays to pass through than the normal adjacent bone. On fluoroscopy screening such an area would look lighter than the surrounding bone, in keeping with the definition of lucent as glowing or luminous, but on radiographs (where most radiolucent lesions or entities are described) this area would look darker. In the context of a well-established convention, this counterintuitive meaning of the word is widely understood and does not cause any day-to-day confusion, but it serves as a reminder of how important a common frame of reference is when making sense of images and how to describe them.

EGYPTIAN ORTHOGONAL VIEWS

The highly distinctive ancient Egyptian schema for depicting the body relies on fusing different characteristic views of the human figure to produce a coherent single image.[35] Figure 1.8 provides a typical example – the feet and legs are shown in profile, or side-on (a lateral view in radiographic terms), as are most of the facial features (nose, mouth, and chin). However, the chest and shoulders, together with the eye, are shown in a frontal view, perpendicular (or "orthogonal") to the legs and

FIGURE 1.8　Sekhmet (Sekhet). Ink drawing. A typical example of stylised ancient Egyptian body representation – the feet and legs are shown in profile, as are most of the facial features (nose, mouth, and chin). However, the chest and shoulders, together with the eye, are shown in a frontal view, perpendicular to the legs and facial features. Credit: Wellcome Collection. Public Domain Mark.

facial features. In this example the position of the arms is a little more complicated – the flexed left arm of the male figure would produce a lateral view of the elbow if X-rayed from our viewpoint, whilst the straight right arm would give a frontal (or "anteroposterior") view.

The Egyptian approach was a deliberate, stylised method to represent the body. We have already seen that Dibutades selected her lover's profile to capture his likeness, and if we consider the shadow of a head as cast side-on rather than front to back (or back-to-front for that matter) it is easy to understand why. Similarly, the profile view of feet and knees are more distinctive compared to their frontal outline, whereas the eye and chest are more amenable to representation from a frontal viewpoint.

The use of perpendicular views of body parts pioneered by the Egyptians has been adopted within radiographic practice, albeit in a more technically minded fashion. For most body parts, and certainly the appendicular skeleton (i.e. arms and hands, legs and feet) it is standard radiographic practice to perform two orthogonal views, most commonly a frontal and a lateral of a bone. It is well established that pathology, such as broken bones, dislocations, and even tumours may be hidden on one view but readily visible on the other, hence the radiology/emergency departments' maxim: "one view only is one view too few".

For particular body parts specific optimal angles and positions may be utilised, although the frontal and lateral views remain the most commonly used projections. In learning to interpret radiographs of specific body structures, radiologists become accustomed to the appearances of specific views. Consequently, I feel more comfortable looking at the standard frontal and lateral views of the knee or elbow than I do looking at oblique (or diagonal) views, although these are used from time to time. My internal representation of the body (at least in radiographic format) is therefore skewed by the characteristic views of body structures most frequently utilised in radiographic practice. The selective use of particular viewpoints, therefore, has the capacity to distort our conceptual appearance of the body in the same way that the Egyptian representation of the human figure appears distorted to modern eyes. Many would likewise argue that the prevalent portrayal of the female figure in Western media in recent decades has had a similar distortive effect on body image.

QUANTISATION

A fundamental mechanism of any digital image-capture technology (including digital radiography, but also essential in ultrasound, nuclear medicine, CT, and MRI) is sampling using a grid-based system and the allocation of an integer value within a finite range to each sampled location.[36] In doing so, any resultant image is constructed from discrete units, which, when displayed on a computer monitor or other screen, are designated as individual picture elements (or the more familiar "pixels" for short). The fragmentation of an image into unitary components has numerous historical precedents, of which we will briefly consider two – Roman mosaics and woven cloth.

ROMAN MOSAICS

While Roman mosaics (including that shown in Figure 1.9) do not conform to a strict grid system in terms of how they are arranged, the use of relatively uniform square (or close to square) tesserae (or tessellae) is a legitimate precursor of the pixel. Within any one part of a design the size of tiles remains of similar size (most commonly 8–12 mm), but as increasingly sophisticated colouring and shading effects were employed, smaller sized tiles (down to around 4 mm) were used to allow more detail, facilitating a more "painterly" style known as opus vermiculatum. A commonly employed scheme used larger sized tesserae around the periphery of the design often showing repeating, geometric-style patterns, with a central panel utilising smaller tiles to enable a detailed motif such as a face or battle scene[37] (see Figure 1.10). The use of different sized tiles for different purposes, or as a budgetary compromise (the smaller the tile, the more intricate and costly the labour per unit area) also has parallels in medical imaging where the size of pixel (or voxel) may need to be adjusted to take account of a variety of factors (such as spatial resolution, radiation dose, scan acquisition time, and so on).

WOVEN CLOTH

The production of woven fabric, particularly that utilising a loom, offers a means of integrating an image or visual motif into the cloth. While embroidered designs stitched into cloth are likely to predate this as a mechanism for decorating clothing and other textiles, our interest in the woven design relates to the "quantised", or unitary fashion by which the image is created. As mosaics are formed from individual, approximately uniform tiles, woven fabric allows an image to be formed from pixel-like configurations of the thread. This can either relate to a short length of thread partitioned by contrasting thread running in parallel on each side and by thread running perpendicularly over the top, or in the case of woven lace or similar the presence or absence of thread at a particular site – akin to the "on/off" signal described by Gombrich.[27]

Textiles are mostly perishable and tend to disintegrate over time. The early history of fabric manufacture is therefore based on patchy evidence, but the origin of weaving is thought to have originated with basketry utilising tender leaves and stems, through to matting, netting, and cordage.[38] Woven woollen fabric found at a Neolithic settlement in Çatal Hüyük, Turkey, has been dated to between 7400 and 6200 BCE, with woven flax found at the Dzudzuana cave in the western part of the Republic of Georgia dating back 19,000 to 23,000 years ago.[39] Indirect evidence of woven fabric in the form of imprints on clay fragments and the possible depiction of woven clothing items including hats in the "Venus" figurines from the Czech Republic dates back a little further to perhaps 27,000 years ago.[40]

Early applications of woven fabric were primarily related to survival, but decorative embellishments soon followed. Evidence of the earliest surviving tapestry weaving (in which the decorative or artistic dimension of the cloth becomes the primary purpose) comes from linen fragments discovered in the tomb of the Egyptian

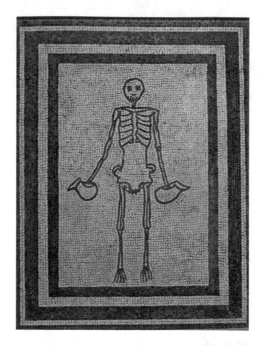

FIGURE 1.9 Photograph of a mosaic from the Roman city of Pompeii representing Death, showing a skeleton of a wine bearer. Image credit: Ruth Swan/ Shutterstock

FIGURE 1.10 Ancient mosaic in Kourion archaeological park near Limassol Cyprus. A woman is portrayed in the central circular portion of the mosaic, with small-sized tesserae used to deliver more detail. Around the periphery, larger tiles are employed in repeating geometric patterns, framing the portrait. Image credit: Alfiya Safuanova / Shutterstock

Pharaoh Thutmose IV, dated between 1483 and 1411 BCE.[41] The Egyptians, in turn, are thought to have acquired the technique of tapestry weaving from the ancient peoples of Mesopotamia. Tapestry was also well established in early civilisations across the globe (including Peru and China)[42] and untangling the threads, so to speak, is a difficult undertaking.

While woven cloth is included here primarily as an example of a precursor of an image composed of quantised or unitary components, we should also make note of aspects related to tapestry conventions. A tapestry is composed of two sets of threads running perpendicularly to one another. The *warp* threads are aligned in a lengthwise or longitudinal axis, whilst the *weft* threads run transversely, or side to side. However, this notation relates to the alignment of thread on the loom. In European tapestries since the middle ages, the design has been executed at right angles to the loom, such that the completed work is hung with the warps running horizontally. By contrast, in Chinese tapestry the warps typically run vertically.[42]

A traditional form of Chinese tapestry known as *kesi* uses fine silk threads, and the weave is finished perfectly on both sides of the fabric, so that the finished work can be displayed on either side. We have already started getting to grips with tonal or greyscale image inversion/reversal. Kesi tapestry introduces us to left/right image reversal, which we will consider in more detail in Chapter 4. The use of fine silk thread (typically around 24 threads per centimetre) in this discipline contrasts with the thick woollen thread popular in European tapestry of the 14th century (The *Angers Apocalypse* tapestry of this era used five threads per centimetre). The calibre of thread diminished over time in Europe reaching 16 threads per centimetre by the 19th century in the French Beauvais factory, before a revival of thicker thread in the 20th century.[41]

The use of thick woollen thread in 14th-century European tapestry may have been informed by practical as well as stylistic considerations – the large-scale pieces commissioned by the nobility are thought to have provided some degree of insulation in otherwise draughty castles or large houses.[42]

In the context of examining how one artistic technique or innovation leads to another, it is instructive to note that the term "cartoon" was originally used to describe the design template from which the tapestry was woven. Likening the Bayeux tapestry to a historical cartoon strip is therefore rather tautological. The computer-generated, pixelated cartoon characters featured in the most popular family movies of recent years can also be said to share some creative heritage with tapestry weaving, but we shall also see in Chapter 8 that medical imaging has been a direct beneficiary of computer animation techniques first developed to bring cartoon characters to life in digital form.

Now take a look at Figure 1.11 which shows an engraving of an 18th-century silk loom and a sample of the floral design on the resultant fabric. This demonstrates a grid-based system and the allocation of particular tonal values to each specified location within the grid. We are really very close – in conceptual terms at least – to a digital image display.

At the beginning of the 19th century, Joseph Marie Jacquard introduced a loom which enabled mass production of textiles with complex woven patterns. The patterns were controlled by pieces of card marked with a series of holes which determined the movements of the loom. The hole-punched cards would later be repurposed in the 1880s and 1890s by Herman Hollerith in the United States to assist in gathering census data. Hollerith used the punched card system with his tabulators – a core product of the company that would eventually become IBM.[43, 44]

LINEAR PERSPECTIVE

During the early 15th century in Renaissance Italy, Filippo Brunelleschi and Leon Battista Alberti developed the technique of linear perspective, partly based on an improved understanding of optics.[45] This led to a number of devices and techniques employed by artists to achieve paintings and drawings in a topographically accurate fashion, such as the one illustrated in a famous woodcut by Albrecht Dürer (Figure 1.12). In the scene, two men are in the process of producing an image of a lute. We see a line of thread attached at one end to the wall behind the artist and at the other to a rod held in position at the edge of the lute by an assistant, who pulls the line taut such that the thread follows a straight line, mimicking the trajectory of a beam of light. The site of attachment on the wall replicates the position of the viewer, where light would converge at the eye. The thread passes through a carefully positioned rectangular wooden frame and the artist records the position that the thread intersects the frame on the canvas or paper which is hinged to swing out of the way while the thread is repositioned. The woodcut shows the image of the lute beginning to take shape as a series of dots – another quantised image – although these were not usually visible in the final drawing or painting.

Another technique which produces a similar result is described by Leonardo da Vinci:

> Obtain a piece of glass as large as a half sheet of royal folio paper and fasten this securely in front of your eyes, that is between your eye and the thing you want to portray. Next position yourself with your eye at a distance of two-thirds of a braccio from the glass and fix your head with a device so that you cannot move it at all. Then close or cover one eye, and with the brush or a piece of finely ground red chalk mark on the glass what you see beyond it.[46]

Alberti himself devised the "reticolato" (also known as Alberti's "grid", "grill", or "graticola"), a grid housed within an otherwise transparent rectangular frame. Such a device is illustrated in Figure 1.13 and is also depicted in another famous Dürer woodcut *Draughtsman Making a Perspective Drawing of a Reclining Woman*, from 1525. A similar device featured prominently in the Peter Greenaway film *The Draughtsman's Contract* (1982).

For each of these methods (thread, pane, or grid) the use of a rectangular plane to define a specific location of the image in space has significant implications. The rectangle becomes cemented as the default shape of representational images, and the choice of viewpoint becomes critical in determining the nature of the image (discussed further in Chapter 5). Unlike the Egyptian approach, the action is portrayed

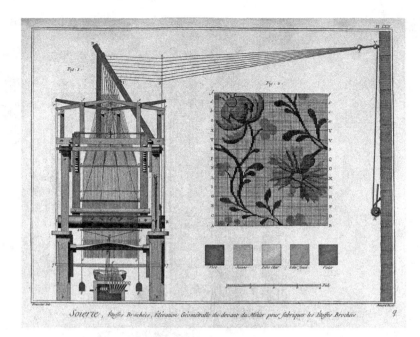

FIGURE 1.11   An engraving of an 18th-century silk loom showing a sample of the floral design on the resultant fabric. Credit: Engraving by R. Benard after L.-J. Goussier. Wellcome Collection. Attribution 4.0 International (CC BY 4.0).

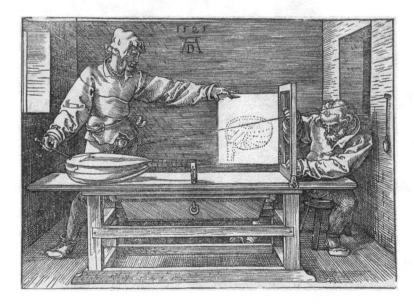

FIGURE 1.12   Woodcut by Albrecht Dürer, showing two men in the process of producing an image of a lute. Image credit: Shutterstock

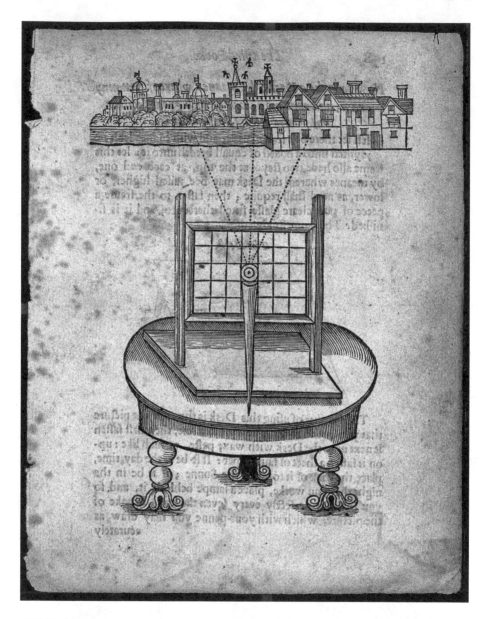

FIGURE 1.13   An engraving showing a circular table upon which a perspective grid has been placed for viewing a town with a river. The grid is housed within an otherwise transparent rectangular frame (also known as Alberti's "grid", "grill", "reticolato", or "graticola"). Credit: Wellcome Collection. Attribution 4.0 International (CC BY 4.0).

from a solitary viewpoint. It is also worth noting that while the optics on which linear perspective relies mirrors that of the lens-based human visual system (discussed in the next chapter), it supposes a monocular situation. Leonardo instructs us to close or cover one eye for the system to work, whilst others have suggested the "reticolato" was of limited use in actually acquiring images due to the difficulty of keeping one eye in exactly the same place.[47] While the intention of linear perspective is to render a more accurate, life-like representation of the world, in doing so it employs a conceptual framework that deviates from how we do actually see.

The perspective devices described above have been characterised as direct forerunners of the digital imaging technique of ray tracing, widely employed in computer-generated image production.[48] As we have already seen, the use of a grid-based system is a fundamental feature of digital image capture and the introduction of the reticolato is certainly an important milestone in this regard. Combining the image acquisition method from Dürer's woodcut (Figure 1.12) and the image display technique used in loom-woven fabric (Figure 1.11), we are not so very far away from reverse-engineering a CT scanner.

IMAGINARY CROSS-SECTION

It could be argued that cross-sectional images have existed in some capacity since the first fruit was sliced in half or the first tree cut down. We shall see in subsequent chapters how anatomical representations of the body developed through the Renaissance and beyond helped to crystallize the conventions of how to display the human frame in planar sections. These conventions would subsequently become adopted in the display and interpretation of cross-sectional imaging techniques such as CT and MRI examinations within the last 50 years.

It will become clear that many methods anatomists employed in displaying the body use a certain degree of artistic license or sleight of hand, and we will see that such images were lacking neither creativity nor imagination. However, a central tenet of Renaissance-era anatomy and beyond was that the images of the body were directly based on the painstaking dissection of the human cadaver – should you have corpses available and sufficient dissection skills you could, in theory at any rate, replicate those images in the flesh.

Before we get to those anatomical representations in Chapters 3, 5, 6, and 8, let us first take a look at Figure 1.14, an illustration from a 19th-century geology text, showing a cross-section of a lead mine in Derbyshire. It depicts a limestone cave found in 1822 by lead miners, in which were discovered the skeletal remains of a horse, a reindeer, an auroch, and a near-complete skeleton of a woolly rhinoceros.[49] Here is an example of the cross-section employed in a fictitious, imaginative fashion.

The view of the mine we enjoy could never be achieved in real life, unlike sliced fruit, felled tree trunks, or dissected corpses. In this sense, this cross-sectional image perhaps shares more conceptual heritage with the planar images of the body delivered by CT and MRI scanners which, although similar in appearance to anatomical sections, could never be directly replicated in a living subject (any "slicing" of the

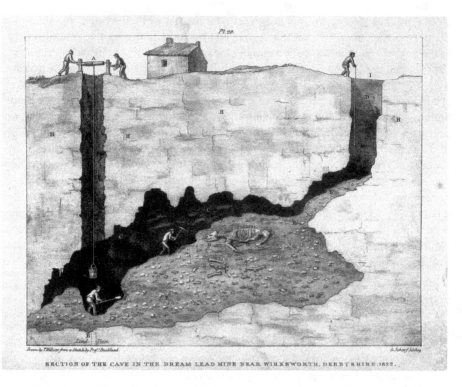

Pl. 20.

SECTION OF THE CAVE IN THE DREAM LEAD MINE NEAR WIRKSWORTH, DERBYSHIRE 1822.

FIGURE 1.14   An illustration from a 19th-century geology text, showing a cross-section of a lead mine in Derbyshire, known as "Dream Cave". Lithograph by T. Webster after a sketch by W. Buckland. Credit: Wellcome Collection. Attribution 4.0 International (CC BY 4.0).

body achieved by a scanner only being figurative in nature). There is no shortage of "fictional" cross-sectional images or diagrams to choose from – you can probably conjure an image of the Earth's internal structure from memory, and diagrams of the eye (as featured in the next chapter) are likely to be a familiar sight. In each case, the subject matter may be real enough, but the method of representation is an abstract construction of the human imagination.

Recent carbon dating of the animal skeletons suggests they were around 45,000 years old[50] (the rhino was a more recent arrival, around 37,000 years old)[51] so were alive roughly the same time as humans were decorating the walls of another limestone cave, located on the other side of the world in Sulawesi, Indonesia. The Derbyshire cavern is variously known as Dream Cave, Dream Mine, or Dream Hole – the common thread being the Dream component. We have already seen that human dreaming and imagination have been a critical driving force of artistic vision and innovation. The journey of apparent technical progress through human history in delivering ever more sophisticated, accurate, and lifelike means of representing the body in visual form has always been in tension with the more nebulous considerations of where human imagination and reality collide.

PLATO'S CAVE

It would therefore feel remiss to conclude this chapter, which has featured caves and shadows so prominently, without recounting Plato's *Allegory of the Cave*. The story is told in the form of a conversation between Socrates and Glaucon in *The Republic*. Socrates describes people who, since childhood, have been shackled in a cave, only able to look forwards to the wall of the cave. Onto this wall are projected the shadows of other people who are able to walk freely on a platform located behind the prisoners, but in front of a fire which casts the shadows. Socrates speculates that in such a situation the prisoners would interpret the shadows as reality itself – "do you not think that they would regard that which they saw on the wall as beings?"[52]

The allegory has been interpreted on multiple different levels, including the search for philosophical truth or religious revelation. An equally applicable take on the story relates to our willingness to accept visual representations as an ersatz reality. In *A History of Pictures* Martin Gayford points out:

To a modern eye, [the prisoners] look a little like the members of an audience watching a film, who are after all, in a dark space, looking at shadows on the wall. Classical scholars have thought the same: F. M. Cornford suggested that the best way to understand what Plato meant was to replace 'the clumsier apparatus' of the cave and the fire with the cinema. Since then, the comparison between Plato's prisoners and viewers of computer screens, film and the entire gamut of visual media has been made time and again.[53]

It is therefore no shocking revelation to suggest that shadows on surfaces, whether artistic or radiographic, have the potential to distort or misrepresent the reality they are intended to depict, discussed further in Chapters 3 and 4.

Plato's cave has also been interpreted as a cautionary tale of the limitations of human sensory perception – we are only aware of a slender portion of reality's rich fabric. Like Dibutades, comforting herself with an outline of her lover's profile in his absence, we are missing out on the true nature of the world as it is. In the next chapter, we will consider some very real limitations of the human visual system, and how the brain compensates for this.

REFERENCES

1. P. Donizetti, *Shadow and Substance*, Pergamon Press, Oxford, 1967, pp 10–13.
2. B.H. Kevles, *Naked to the Bone: Medical Imaging in the Twentieth Century*, Rutgers University Press, New Brunswick, 1997, p 20.
3. W.F. Bynum, *Science and the Practice of Medicine in the Nineteenth Century*, Cambridge University Press, Cambridge, 1994, p 173.
4. *Radiograph Performed by Dawson Turner*, Royal College of Physicians, Edinburgh, 1896.
5. J. Anderson, William Morgan, and X-rays, *Transactions of the Faculty of Actuaries*, 17, 1945, 219–221. doi:10.1017/S0071368600003001
6. A.M.K. Thomas and A.K. Banerjee, *The History of Radiology*, Oxford University Press, Oxford, 2013, p 1.

7. Ibid p 2
8. Iwona Sudoł-Szopińska and Marta Panas-Goworska, Pioneers – The history of musculoskeletal radiology: Ivan Pulyui, www.essr.org/content-essr/uploads/2019/02/ESSR_Pub_Ivan-Pulyui.pdf
9. N.O. Virchenko, 170th anniversary of the birth of Ivan Pulyuy, National Technical University of Ukraine, https://kpi.ua/en/puliuy, last accessed 21/07/21.
10. Wikipedia, Ivan Puluj https://en.wikipedia.org/wiki/Ivan_Puluj, last accessed 21/07/21.
11. Ivan Pulyuy, Ternopil National Technical University of Technology, http://tntu.edu.ua/?print=uk/news/1320, last accessed 21/07/21.
12. D. Kulynyak, NOTEWORTHY UKRAINIANS: Ivan Pului, the discoverer of X-rays, www.ukrweekly.com/old/archive/2000/280012.shtml, last accessed 07/09/20.
13. Lviv Today, The Ukrainian inventor of the X-ray, *Lviv Today*, Issue 83, October 2015, http://www.lvivtoday.com.ua/lviv-history/4370, last accessed 21/07/21.
14. E. Těšínská, Johann Puluj (1845–1918): His career and the "invisible cathode rays", *IUCr Newsletter*, 28(2), 2020, www.iucr.org/news/newsletter/etc/articles?issue=148371&result_138339_result_page=26, last accessed 21/07/21
15. W. Savchuk, The naturalist I. P. Puljuj and the discovery of X-rays, *The Global and the Local: The History of Science and the Cultural Integration of Europe. Proceedings of the 2nd ICESHS (Cracow, Poland, September 6–9, 2006)*, M. Kokowski (ed), http://www.2iceshs.cyfronet.pl/2ICESHS_Proceedings/Chapter_10/R-2_Savchuk.pdf
16. P. Belluschi, Science and the modern architect, in *The New Landscape in Art and Science*, G. Kepes (ed), 3rd edition, Paul Theobald & Co, Chicago, 1963, p 28.
17. M. Aubert, R. Lebe, A.A. Oktaviana et al. Earliest hunting scene in prehistoric art, *Nature*, 576, 2019, 442–445. doi:10.1038/s41586-019-1806-y
18. Hand paintings and symbols in rock art, Bradshaw Foundation, www.bradshawfoundation.com/hands/, last accessed 21/07/21
19. D.L. Hoffmann, C.D. Standish, M. García-Diez, P.B. Pettitt, J.A. Milton, J. Zilhão, J.J. Alcolea-González, P. Cantalejo-Duarte, H. Collado, R. de Balbín, M. Lorblanchet, J. Ramos-Muñoz, G.-Ch. Weniger and A.W.G. Pike, U-Th dating of carbonate crusts reveals Neandertal origin of Iberian cave art, *Science* 359(6378), 2018, 912–915. doi:10.1126/science.aap7778. PMID 29472483.
20. Prehistoric Colour Palette, Visual arts cork, http://www.visual-arts-cork.com/artist-paints/prehistoric-colour-palette.htm, last accessed 21/07/21.
21. D. Snow, Sexual dimorphism in European upper Paleolithic cave art, *American Antiquity*, 78(4), 2013, 746–761. doi:10.2307/43184971
22. J. Wagelie, X-ray style in Arnhem Land Rock Art, in *Heilbrunn Timeline of Art History*, The Metropolitan Museum of Art, New York, 2000, http://www.metmuseum.org/toah/hd/xray/hd_xray.htm, last accessed October 2002.
23. Excerpt from Pliny, *Natural History*, Book XXXV, https://pages.wustl.edu/westernsummer2017/reading-pliny
24. Pablo Garcia, The origin of painting, projection systems, https://projectionsystems.wordpress.com/2009/09/06/the-origin-of-painting/, last accessed 21/07/21.
25. Athenagoras [Embassy, 17] as quoted in William J. Mitchell, *The Reconfigured Eye: Visual Truth in the Post-Photographic Era*, MIT Press, Cambridge, 1992, p 227.
26. R. Rosenblum, The origin of painting: A problem in the iconography of romantic classicism, *The Art Bulletin* 39(4), 1957, 279–290.
27. E.H. Gombrich, *Art and Illusion: A Study in the Psychology of Pictorial Representation*. 5th edition, Phaidon Press, London, 1977, pp 34–35.
28. J. Boardman, *The History of Greek Vases: Potters, Painters and Pictures*, Thames & Hudson, London, 2001.
29. A. Thomas, Via email correspondence.

30. R.H. Wachsberg, Inversion of the grayscale display to facilitate viewing of computed tomographic scans by sonographers, *Ultrasound Q.* 24(3), 2008, 179–180. doi:10.1097/RUQ.0b013e3181836b8c

31. C. Xia, L. Xu, B. Xue, F. Sheng, Y. Qiu and Z. Zhu, Grayscale inversion view can improve the reliability for measuring Proximal Junctional Kyphosis in Adolescent Idiopathic Scoliosis, *World Neurosurg.* 119, 2018, e631–e637. doi:10.1016/j.wneu.2018.07.224

32. W. Sun, J. Zhou, X. Qin, et al. Grayscale inversion radiographic view provided improved intra- and inter-observer reliabilities in measuring spinopelvic parameters in asymptomatic adult population, *BMC Musculoskelet Disord.* 17(1),2016, 411. Published 2016 Oct 3. doi:10.1186/s12891-016-1269-3

33. M. E. Sheline, I. Brikman, D.M. Epstein, J.L. Mezrich, H.L. Kundel and R.L. Arenson, The diagnosis of pulmonary nodules: Comparison between standard and inverse digitized images and conventional chest radiographs, *AJR Am J Roentgenol.* 152(2), 1989, 261–263. doi:10.2214/ajr.152.2.261

34. J. Kirchner, D. Gadek, J.P. Goltz, et al. Standard versus inverted digital luminescence radiography in detecting pulmonary nodules: A ROC analysis, *Eur J Radiol.* 82(10), 2013, 1799–1803. doi:10.1016/j.ejrad.2013.05.001

35. E.H. Gombrich, *The Story of Art.* 15th edition, Phaidon, 1989, p 35.

36. J.M. William, *The Reconfigured Eye: Visual Truth in the Post-Photographic Era*, MIT Press, Cambridge, 1992, p 60.

37. M. Cartwright, Roman Mosaics https://www.ancient.eu/article/498/roman-mosaics/

38. K. St Clair, *The Golden Thread: How Fabric Changed History*, John Murray, London 2018.

39. Ibid [loc 471]

40. O. Soffer, J.M. Adovasio and D.C. Hyland, The "Venus" Figurines: Textiles, basketry, gender, and status in the Upper Paleolithic, *Current Anthropology*, 41(4), August–October, 2000, 511–537

41. M. Jarry, Tapestry, *Encyclopedia Britannica*, 8 March 2019, https://www.britannica.com/art/tapestry, last accessed 21 July 2021.

42. C. Missaggia, European Tapestries: History, Conservation, and Creation, Senior Honors Projects. Paper 313, 2013.http://digitalcommons.uri.edu/srhonorsprog/313http://digitalcommons.uri.edu/srhonorsprog/313

43. K. St Clair, *The Golden Thread: How Fabric Changed History*, John Murray, 2018 [loc 244].

44. The IBM punched card, IBM, www.ibm.com/ibm/history/ibm100/us/en/icons/punchcard/, last accessed 21 July 2021.

45. D. Hockney and M. Gayford, *A History of Pictures*, Thames & Hudson, London, 2016, pp 94–96.

46. Leonardo Da Vinci as quoted in William J. Mitchell, *The Reconfigured Eye: Visual Truth in the Post-Photographic Era*, MIT Press, 1992, p 154.

47. B. Smith, T. Grid, in D. Montello (ed.), *Spatial Information Theory. Foundations of Geographic Information Science* (Lecture Notes in Computer Science 2205), Springer, Berlin/New York, 14–27.

48. J.M. William, *The Reconfigured Eye: Visual Truth in the Post-Photographic Era*, MIT Press, Cambridge,1992, pp 153–6.

49. K. Vahedv, The secrets of Wirksworth's Dream Cave, *Derbyshire Life*, 18/01/19, www.derbyshirelife.co.uk/out-about/places/wirksworth-dream-cave-1-5849057, last accessed 21 July 2021.

50. D.A. McFarlane, J. Lundberg, G.V. Rentergem, E. Howlett and C. Stimpson, A new radiometric date and assessment of the Last Glacial megafauna of Dream Cave, Derbyshire, UK, *Cave Karst Sci.* 43(3), 2016, 109–116.

51. D. McFarlane, J. Lundberg and D. Ford, The Age of the Woolly Rhino from Dream Cave, Derbyshire, UK, *Cave Karst Sci.* 27(1), April 2000, pp 25–28.

52. Plato, *The Allegory of the Cave*, Republic, VII, 514 a, 2 to 517 a, 7, Translation by Thomas Sheehan, https://web.stanford.edu/class/ihum40/cave.pdf

53. D. Hockney and M. Gayford, *A History of Pictures*, Thames & Hudson, London, 2016, p 76.

2 The Eye's Mind

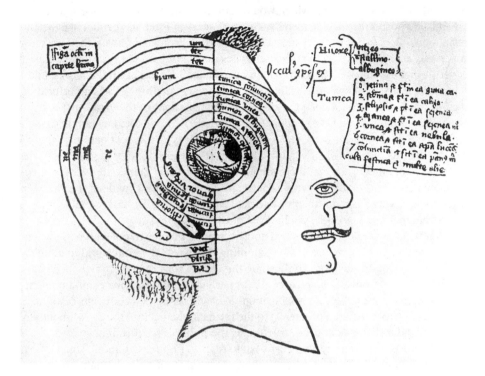

Schematic Eye illustration from a 14th-century manuscript in the British Museum. Credit: Wellcome Collection. Attribution 4.0 International (CC BY 4.0)

This chapter will review various aspects of human visual physiology salient to how we perceive images, whether radiological or artistic in nature. While starting with the eyes, it will become evident that the brain is largely responsible for how we visualise our external world. A flavour of the complexity of cerebral mechanisms related to vision will be provided by considering several phenomena/conditions in the second half of the chapter, dipping our toes in the vast ocean of visual neuroscience and psychology. For the purposes of this discussion, however, the means by which the brain generates images (whether directly perceived from the external environment, or generated internally in the "mind's eye") is largely going to be considered a "black box". Likewise, for brevity, I have also largely omitted the complexities of colour vision.

DOI: 10.1201/9780367855567-2

THE SCRATCHED VISOR

During the COVID-19 pandemic, I was asked to do some portable ultrasound examinations on patients in the emergency department at the Royal Hospital for Sick Children in Edinburgh. As part of the personal protective equipment, I was required to wear a face visor, which would soon become a commonplace sight in the UK outside hospitals, such as members of staff serving food and drinks in pubs and restaurants. In contrast to the disposable face mask, apron, and gloves, the face visor was a reusable item and had various scratches and marks on the transparent plastic cover from previous use. On putting it on I was acutely aware of all the marks, and I remember thinking "how am I going to be able to see anything?" when performing the ultrasound scan. Yet, once I got going the imperfections of the visor surface were largely forgotten and, as far as I could tell, the visor did not make any tangible detriment to the quality of the scan. In a similar vein, as I type this text onto my laptop I can see the text nice and clearly, but if I look up from the computer to gaze out of the window, I become aware that there are all sorts of grubby marks all over my spectacles which could really do with a thorough wipe. In both situations, I have been able to selectively filter out the noise to achieve a meaningful, interpretable image, something the human visual system is extraordinarily good at.

Even without scratched visors and poorly cleaned glasses to contend with, whenever we open our eyes we are always sampling the external visual world in a highly incomplete, selective fashion. To illustrate this, let us start by travelling with a beam of light reflected from an object, artwork, or perhaps emitted from a computer monitor displaying a radiological investigation. We will imagine this beam of light to pass directly from the area of interest to the retina at the back of the eye without any deviation and will consider each structure as the light passes through.

As we make this journey, you may wish to refer to Figures 2.1 and 2.2, an Optical Coherence Tomography montage showing cross-section of a human eye and a labelled line drawing of the eye in similar cross-section, respectively. On the line drawing, the trajectory of our beam is similar to that of the "pupil" label line, heading to the fovea.

THE CORNEA AND AQUEOUS HUMOUR

The first entity light encounters as it strikes the visual system of an observing human is a thin layer of tears on the surface of the eye. Although considered primarily a protective mechanism to prevent debris and dirt causing damage, the transition in refractive index between the air and the watery solution of tears means that this thin film has a substantial effect in focusing light towards the retina. Beyond the layer of tears, light will pass into the cornea which bulges out as a focal convex protrusion from the rest of the eyeball. Being of similar flexibility and durability to plastic provides a physical barrier protecting the internal contents of the eye. The combination of the curvature of the cornea and the change in density from air to water at the surface of the eye accounts for around 65% of the eye's refractive power (~40 dioptres).[1]

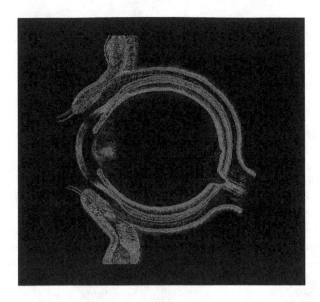

FIGURE 2.1   An Optical Coherence Tomography montage showing cross-section of a human eye, by Jon Brett. Credit: Wellcome Collection. Attribution-NonCommercial 4.0 International (CC BY-NC 4.0).

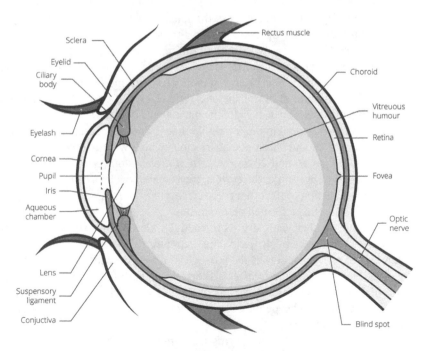

FIGURE 2.2   A labelled line drawing of the eye in cross-section. Diagram by Alexander_P. Image credit: Shutterstock

Light that strikes the cornea perpendicular to its surface is not refracted, so our beam of light does not deviate from its original course, but the cornea also serves to filter out ultraviolet light so absorbs much of the UV component as it passes through. It is worth noting that whilst in humans it is beneficial to filter out UV radiation which is potentially damaging to the lens and retina,[2] many other species make use of the ultraviolet spectrum to visualise their environment. The cornea, therefore, represents a first indicator of the selective nature of what information is utilised and which is discarded.

Once past the cornea, light travels through the anterior and posterior chambers, located between the cornea and iris, and the iris and lens respectively, and for the purposes of this discussion largely indistinguishable from one another apart from the anatomical boundaries. These chambers are composed of aqueous humour, a clear fluid produced continually (albeit at a variable rate) by the ciliary body, a structure located just behind the outer edge of the iris. The fluid serves the function of a blood surrogate in providing nutrition and helping to eliminate waste products in the absence of blood vessels in this part of the eye. There is a continuous cycle of production and drainage, such that the fluid is replaced every couple of hours or so. Issues with drainage of the fluid can result in a harmful build-up of intraocular pressure, leading to glaucoma.[3]

THE LENS

The next structure light will encounter is the lens, but to reach it light first passes through the pupil, a circular aperture within the aqueous humour, the calibre of which is determined by the iris. Maximal dilatation of the iris provides a pupil size 15 times larger than when constricted.[4] The lens itself is a biconvex encapsulated structure with a high content of crystallin proteins (accounting for around 60% of its mass) which offer a high level of transparency and refractility. At least three varieties of crystallins are found in varying concentrations in different parts of the lens providing regional differences in the refractive index. These regional variations help correct for the spherical and chromatic aberrations which would otherwise occur in a homogeneous lens. At around 20 dioptres, the refractive power of the lens is only half that of the cornea, but unlike the cornea, the shape of the lens can be adjusted by contraction and relaxation of the ciliary muscle, and so the lens serves an important role "fine-tuning" the image formed on the retina.

Whilst in youth the lens can vary its refractive index by up to 12 dioptres, this typically falls to 6 dioptres by 40 years and 1 dioptre or less by 60 years. In addition to the loss of refractive power, the passage of time also causes the lens to lose transparency. This predominantly affects short wavelength light, which is increasingly absorbed, giving the lens a slight amber tinge in advancing years and reducing blue light sensitivity in older people. Cataract formation represents a more severe manifestation of this process.

Notably, the surface of the lens forms an impenetrable barrier to cells of the immune system, making the internal contents of the lens an "immunologically sequestered" environment, presumably to prevent any inflammatory response damaging the delicate crystalline structure. A consequence of this splendid isolation is that it retains all the cells produced throughout its life – a unique feature in the body.[1]

VITREOUS HUMOUR

Behind the lens is the vitreous humour (also known as vitreous body), composed of a transparent, colourless hydrogel, and accounting for 80% of the volume of the eye. The vitreous humour serves as a mechanical shock absorber for the eye, absorbing impacts and protecting the lens and retina.[5] Unlike the continuous production and drainage of aqueous humour, vitreous humour is never replenished. Any debris, such as blood cells or other cellular material, which ends up within the vitreous will stay there for good. These foci of detritus can block light passing through the eye, causing the phenomenon of myodesopsia – more commonly referred to as "floaters", and also known as *muscae volitantes* (Latin for "hovering flies"). The gelatinous consistency becomes progressively more liquified with age, which contributes to a higher likelihood of such floaters and a higher risk of eye-related structural issues such as retinal detachment.[5, 6]

Once past the vitreous the light reaches the back of the eye, and the business end of things – the retina. However, the cumulative effect of passing through the cornea, aqueous humour, the lens and the vitreous humour is that less than half the light incident on the surface of the eye actually reaches the retina, the rest being lost to scatter or absorption.[7]

PHOTORECEPTOR CELLS

The ultimate destination for our beam of light is photopigment contained within retinal photoreceptor cells, where energy of the light will be transformed into neural signals. We haven't quite got there yet, and before we do it is worth getting an overview of the composition of the retina.

The photoreceptive layer is composed of two types of cell known as rods and cones. Rods are more sensitive than cones and provide the main mechanism for vision in low-intensity light such as at night. They are named after the relatively uniform, cylindrical shape of the segment of the cell containing the photopigments. Rods are only of one type, which results in monochromatic vision. By contrast, cones come in three varieties, responding optimally to three different (if overlapping) ranges of wavelength. They are named after the tapered, conical morphology of the photopigment segment. It is through the sensory data provided by cones that we are able to see colour and appreciate fine detail, but they require higher levels of illumination, typically requiring around 100 photons of light to activate the photopigments whereas rods can respond to a solitary photon.[8]

Figure 2.3 shows a transverse section of the retina, demonstrating a mixture of rod and cone cells. When I first looked at this image I had assumed the larger cells were rods and the smaller, more numerous cells were cones on the basis that cones offer superior spatial resolution, but in fact the rods are the smaller calibre cells. We will address this apparent anomaly shortly.

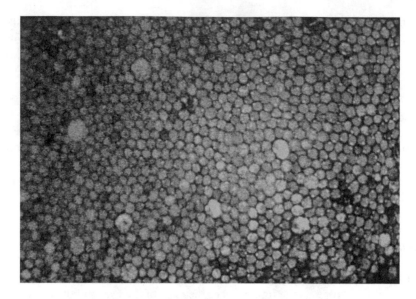

FIGURE 2.3 Microscopic image showing a transverse section of the retina, demonstrating a mixture of rod and cone cells. Credit: Chris Guerin. Wellcome Collection. Attribution-NonCommercial 4.0 International (CC BY-NC 4.0).

In modern (and particularly urban) life, we tend not to make as much use of our rods as our ancestors, and most of the visualisation considered in this text relates to that achieved by cones. Nevertheless, it is interesting to note that rods vastly outnumber cones, with on average 92 million rods compared to 4.6 million cones in the human retina[1] (some sources put the numbers higher at 120 million and 8 million respectively[9]). Although the photopigment in rods is sensitive enough to be stimulated by a single photon, around half a dozen such activations would be required to experience a short dim flash. The sensitivity offered by rods enables, under ideal circumstances, a single candle to be visible from 20 miles away.[7] Faint visibility of the Andromeda galaxy with the unaided eye relates to photons from some 2.537 million light years away (1.492×10^{19} miles).[10]

INVERTED RETINAL STRUCTURE

Back to our more modest journey across the eyeball, light has reached the retina, but is not quite at journey's end. Compared to the retina of a squid or octopus, vertebrate (including human) retinas appear to be constructed "back to front" or in an inverted configuration. In the eyes of a squid (which are otherwise very similar in structure to our own), the light detecting photopigments of the photoreceptors are arranged at the front of the retinal surface (i.e. lens side rather than brain side), with the remainder of the cell structures and the connecting neurons located behind. In the human retina the photopigments in the receptor cells are located close to the posterior wall of the retina, and with the nucleus of each cell located at the front, such that light needs to pass through the cellular machinery before it reaches the photopigments.

In addition to this, there is an extensive layer of other neurons (known as the plexiform) located in front of the photoreceptors which relay the information to the brain (the equivalent cells and axons are located at the back of the retina in the squid, where they don't get in the way). The light also needs to pass through this layer prior to getting to the photopigments (although there is an exception, discussed shortly). These overlying neural structures are largely transparent and do not have much impact on optical quality. Indeed, some research suggests the arrangement of the plexiform neural layer, with cells packed tightly in parallel alongside one another along the axis of incident light, has a beneficial effect by helping to reduce scattered light and hence eliminating noise.[11] This is a similar mechanism to anti-scatter grids used in plain radiography, discussed in Chapter 3.

However, the blood vessels supplying the retina are also distributed over its anterior (lens side) surface, the larger branches of which occupy a significant fraction of the retinal surface, and are dense enough to cast shadows onto the receptor cells. Under normal viewing conditions, we are unaware of these vessels – our brains adapt to the static pattern of branching linear shadows cast onto the retina in early life and edit them out of our conscious perception, through broadly similar mechanisms by which I was able to filter out the scratches on my visor or the smudges on my glasses.[12] On occasion we do become aware of this appearance – it can be seen by shining the beam of a small bright light through the pupil from an extreme angle at the periphery of an individual's visual field (most commonly encountered when eyes are examined with an ophthalmoscope). This results in an image of the vessels being focused on the periphery of the retina, involving portions unadapted to the vascular pattern,[12] known as Purkinje shadows (named after Jan Evangelista Purkinje, the Czech physiologist and neuroanatomist who first described this phenomenon in detail).[13]

The difference in retinal architecture between squids and humans relates to the respective embryological development of the eye originating from different evolutionary history. In the squid both the lens and retina form from skin-forming tissue (surface ectoderm), whereas in vertebrates the retina develops as an eye cup formed by brain-forming tissue (neural crest), but with the lens (and cornea) developing from the surface ectoderm.[14] The interaction between these two distinctive types of tissue in vertebrate development, involving dynamic folding and invagination of the primitive eye structures produces this characteristic retinal morphology,[15] but a full explanation is beyond the scope of this discussion.

A further quirk of the "back-to-front" vertebrate retina is the blind spot. The axons conveying the sensory data from the photoreceptor cells traverse over the front surface of the retina to congregate at the optic disc, a circular structure approximately 1.5 mm in diameter located nasal to the macula (see Figure 2.4). The optic disc forms the first part of the optic nerve, which conveys sensory data from the eye to the brain. The blood vessels that supply the retina course through the optic nerve and enter the eye at the site of the optic disc. No photoreceptors are present at the optic disc, so no visual information is acquired at this site on the retina. Using various manoeuvres you can demonstrate the presence of your own blind spot (an internet search is suggested if interested), but we are not usually aware of it due to two mechanisms. First, the relative positions of the optic discs in each eye enable receptors in one eye to

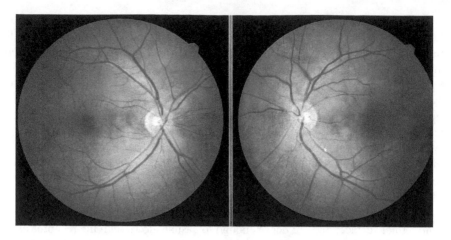

FIGURE 2.4 Fundoscopic photographs of the author's retinas. Image provided by the author.

register what is missed in the blind spot of the other eye. Second, the visual system fills in this region with appropriate sensory information, in a similar manner to that of the shadows of the blood vessels, known as perceptual interpolation.[16]

THE FOVEA

The beam of light we are following is headed for the region of the retina with the highest visual acuity – the fovea centralis, located within the central region of the retina called the macula. Within the fovea, we are heading for the most central portion which provides the most detailed vision, the foveola. Out of a total retinal surface of 25 cm², the fovea represents an area of only 1.5 mm diameter, and the foveola is only 0.3 mm in diameter.[17] However, despite the small size around 75% of light entering the eye is focused onto the fovea.[8]

Outside of the macula, in the periphery of the retinal surface the photoreceptive layer is composed almost exclusively of rods, whereas the foveola is only composed of cones. The remainder of the fovea and the macula are largely composed of cones but with an increasing proportion of rods seen progressively towards the outer edge of the macula.

The fovea, and the foveola, in particular, have the most densely packed receptor cells, but despite a higher number of receptors per unit area (reaching an average density of almost 200,000 per mm²)[1], the retinal layer is actually thinner at this site compared to the rest of the retina. Figure 2.5 shows a microscopic section through the retina, centred at the foveola. The cone receptor cells resemble cress growing in a planting tray – these are particularly tightly bunched with the stalks (actually the segments containing light-sensitive pigment) difficult to discriminate from separate cells, whereas some degree of separation can be seen away from the centre on each side. The reason the fovea is thinner relates to cellular structures lying superficial to

FIGURE 2.5   A microscopic section through the retina, centred at the foveola. Credit: Chris Guerin. Wellcome Collection. Attribution-NonCommercial 4.0 International (CC BY-NC 4.0).

the receptor layer. As we have seen, elsewhere in the retina a layer of neural cells overlies the receptor cells – any light reaching the receptor cells beneath must pass through these neurons before getting to the light-sensitive receptors. At the foveola, the neurons of the plexiform layer are largely displaced to the side, producing the appearance of the dip or valley in Figure 2.5, known as the foveal pit, also seen in Figures 2.1 and 2.2. The foveola is also free of blood vessels (including capillaries), and larger calibre vessels are distributed around the periphery of the macula to enable the most unobscured view for central, high acuity vision ("the foveal avascular zone").[16] Relative absence of the blood vessels around the site of the macula can be appreciated in the retinal photographs of Figure 2.4.

One estimate suggests up to a quarter of light directed onto the fovea (almost 20% of all light entering the eye) is focused onto a region of the foveola containing only 30 or so cone cells.[8]

Another paper draws attention to the apparent low resolution offered by the human fovea, suggesting a resolving power of only 0.25 megapixels.[16] By comparison my own smartphone, already a few years out of date at the time of writing, has a 12.3-megapixel camera.

THE PLEXIFORM LAYER

The process of light stimulating photopigments located within rods and cones leads to electrical activity conveying this visual data to the brain via the axons in the optic nerve. However, this process is conducted via several intermediary steps, each of which is subject to modulation from other neural processes. A huge quantity of information is integrated and compressed within the retinal circuits; input from around 100 million receptor cells converges on around 1 million ganglion cells, the axons of which form the optic nerve. This means each ganglion cell integrates information from an average of 100 photoreceptor cells. The level of integration varies from around one to five cones at the foveola (maintaining a high level of acuity/spatial resolution) to around a thousand rods from peripheral areas of the retina. It is this pooling of data from a large number of rods which accounts for the lower spatial resolution compared to cones, despite individual rods being smaller in calibre than cones.[1]

Relaying of information from the photoreceptors to the ganglion cells is performed by the bipolar cell, but bipolar cells are modulated by two other types of cells found in the plexiform layer: horizontal cells and amacrine cells. Horizontal cells modulate the transfer of information from the photoreceptor cell to the bipolar cell and amacrine cells influence the data flow at the junction of the bipolar cell with the ganglion cell. Adjacent photoreceptor cells are also in communication with one another via gap junctions.[8]

The summation, integration, and modulation of data from the photoreceptor cells mean the data conveyed via the optic nerve is very much an "edited highlights package" of the visual information incident on the retina at any one time, in which noise is filtered out and certain features enhanced.

LATERAL INHIBITION/MACH BANDS

Lateral inhibition describes a mechanism by which a neuron's response to a stimulus is inhibited by the excitation of a neighbouring neuron. It is found in all animal species with some level of complexity in their nervous system and was first described in the compound eye of the horseshoe crab in the 1950s.[18] While our interest is in relation to vision, lateral inhibition is also utilised in the processing of sensory data in the olfactory, tactile, and auditory systems.

In visual perception, the purpose of lateral inhibition is to facilitate edge detection. In the human visual system this is performed both within the retina, and then reinforced within the brain via visual processing in a location known as the lateral geniculate nucleus. In the retina this is achieved by the grouping of a cluster of adjacent photoreceptor cells into a functional unit known as a concentric receptive field, arranged as a central circular area (centre) surrounded by an outer ring (surround). Light incident on the two regions of the receptive field has opposite effects, and so the centre and surround are said to be *antagonistic* to one another (centre-surround antagonism). When the intensity of light falling in the "on" region increases or decreases, the strength of the cell's response changes in the same direction. An increase or decrease in light intensity in the "off" region causes the response to change in the opposite direction. The output of each receptive field is relayed to the optic nerve via a ganglion cell. The visual field is composed of a mixture of "on" and "off" ganglion cells which produce opposite responses.[2]

Lateral inhibition is responsible for the visual illusion known as Mach bands, named after physicist Ernst Mach who described the effect in 1865. He observed that when two bars of uniform tone, one dark and one bright, are placed next to each other, the eye perceives bands of increased darkness at the edge of the dark band and increased lightness at the edge of the light bar. These bands do not exist but are an illusion caused by lateral inhibition via our centre-surround receptive fields.[19, 20]

Mach bands are illustrated in Figure 2.6 – in the left-hand collection of grey rectangles you should perceive increased darkness and lightness respectively at the junction of two adjacent rectangles as you work from top to bottom. However, the right-hand column rectangles are of the same tone as the corresponding left-hand ones but separated by white bars which negate the effect of lateral inhibition.

Mach effect is of importance within radiology, particularly in the interpretation of plain radiographs (or X-rays), where it is possible to both be fooled by it (perhaps interpreting a Mach line as a fracture or other pathology that is not really there – a false positive finding), or alternatively using the effect to dismiss appearances which really are present (a false negative report).

FIGURE 2.6   Illustration of Mach bands. In the left-hand column, rectangles of grey of differing shades are located immediately adjacent to one another – you should perceive increased darkness and lightness, respectively, at the junction of two adjacent rectangles as you work from top to bottom. However, in the right-hand column rectangles of the same tone as the corresponding left-hand ones are separated by white bars, which negates the effect of lateral inhibition, such that each rectangle is perceived as being of uniform tone. Diagram by Zhitkov Boris. Image credit: Shutterstock

EYE MOVEMENTS

While in totality, human binocular vision covers a horizontal range of around 150 degrees, the visual angle covered by the fovea is only about 2 degrees, equivalent to the size of your thumbnail held at arm's length. This small patch offers the most detailed colour representation and the best spatial resolution.[9] At any moment in time the human eye samples both this small central (or foveal) field with high acuity and a much larger segment of the optic array (the peripheral field) with low acuity.[21] This model of central high focus and rather fuzzy periphery was dramatically demonstrated to me when mountain biking not so long ago, with intense visualisation of a small patch of the path around two metres ahead of my front wheel (looking for potential hazards like rocks and so forth) and everything around the outside of this being rather blurry. Maintaining eye contact in close proximity represents another manifestation of this effect – you have to pick one eye or the other to focus on (or alternate between the two), rather than being able to see both in focus simultaneously.

To achieve sampling of the visual field, the eyes are in a state of near-perpetual motion via four main types of characteristic ocular movement: saccades, smooth pursuit movements, vergence movements, and vestibulo-ocular movements. In addition to these controlled (if not always voluntary) movements, the eyes are also subject to a continuous background tremor. This results from instability in the three pairs of antagonistic extraocular muscles which hold the eyes in position. Consequently, the image on the retina is in constant motion, any point on it moving by about the distance between two adjacent foveal cones in 0.1 second.[21]

Saccadic eye movement describes rapid, jerky movements which abruptly change the point of fixation and help localise a particular object or scene onto the fovea. Saccades occur up to four times per second and can reach speeds of up to 500 degrees per second.[17] Although jerky, these movements are not imprecise, with an accuracy of 0.2 degrees or better.[22] However, they are sometimes described as "ballistic", as once initiated a saccade cannot be terminated, rather requiring an additional corrective saccade to enable repositioning.[23] While reading this sentence your eyes will probably make about half a dozen jerky saccades rather than following the line smoothly.[22]

Saccadic eye movement is used as the basis of optical illusions such as that shown in Figure 2.7, heavily influenced by the "op-art" of Bridget Riley, in which the concentric zig-zags appear to move as our eyes survey the image.

The process of scanning a scene with saccadic eye movements is a predominantly involuntary and unconscious process but can be likened to the large-scale multidirectional eye movements we are familiar with when driving a car. We look through the front windscreen to see where we are going, but our focus oscillates between numerous features in a variety of positions (the road itself; other vehicles; pedestrians on the pavement; road signs, and traffic signals), we also divert our gaze to the rear-view mirror, and to each of the wing mirrors. From time to time we look down to check the speedometer, or the fuel gauge and on occasion down and to the side to fiddle with the car stereo. Each of these separate views (akin to different camera shots in a movie scene) contribute to our moment-to-moment conscious

FIGURE 2.7 Optical illusion, heavily influenced by the "op-art" of Bridget Riley, in which the concentric zig-zags appear to move as our eyes survey the image. Image credit: Betacam-SP/ Shutterstock

experience of driving in a manner analogous to each fragmentary image of portions of the visual field being knitted together to produce a fluent and coherent visual consciousness.

In contrast to the darting movements of saccades, once a moving object or target is fixated, much slower smooth pursuit movements are employed to maintain it within foveal vision (foveation) as it moves, or as the observer moves.[21] An interesting feature is that while most of us can achieve smooth pursuit movement effortlessly whilst actually following a mobile object, only highly trained observers can make a smooth pursuit movement in the absence of a moving target, the rest of us resorting to saccadic movements.[23] I have, as yet, been unable to establish what the purpose of training to follow imaginary moving targets actually is.

In charting the passage of a beam of light through a solitary eyeball I have so far largely ignored the binocular nature of human vision, but vergence movements are specifically required to ensure both eyes are directed at the object of interest. In contrast to the other eye movements, where both eyes move in the same direction (known as conjugate eye movements), vergence movements result in each eye moving in opposite directions (disconjugate or disjunctive movement), either converging towards one another to enable focusing on a nearby object of interest, or diverging away from each other to visualise more distant features. Vergence movement is smooth and continuous, similar to pursuit movement, rather than jerky saccades.[21] The difference in the view between each eye enables depth perception.[22]

Vestibulo-ocular movements are reflexive responses which stabilise the position of the eyes relative to the external world during head movements.[23] Receptors in the semicircular canals, located in the inner ear, respond to active or passive rotational (angular) accelerations of the head. Body rotation is matched by corrective counter-rotation of the eyes such that gaze direction is unaltered, maintaining the image in the same position on the retina, enabling good quality vision to be preserved.[22]

Despite these corrective mechanisms, the cumulative effect of all these eye movements is that while the eyes are capable of highly precise tracking of objects, and speedy coverage of a wide visual field, the sensory input is essentially discontinuous, and frequently corrupted by noise. Blurred, poor quality images during movements are interspersed with static, high-resolution images when the eyes are (relatively) stationary. The data stream from which our visual experience is constructed is, therefore, composed of varying frame rates and varying image quality.[17] We also become aware of different visual characteristics at different times, with the perception of colour occurring 40 milliseconds before that of form, and 80 milliseconds before the perception of motion.[24] Despite this our conscious perception tends to be fluid and immersive.[17]

VISUAL DATA PROCESSING

We have already seen that in relaying data from the retinal photoreceptors to the optic nerve there is a pooling of data (100 million photoreceptors → 1 million ganglion cells). Quantifying this data bottleneck for both eyes is estimated to be around 10 billion bits of information arriving at the retina each second being compressed into 6 million bits entering into the optic nerves.[17]

The sensory visual information conducted by the optic nerves is relayed to multiple regions of the brain, with an increasingly widespread distribution of this visual information throughout the cerebral hemispheres being revealed by neuroimaging studies such as functional MRI. However, the lion's share of the information from the retina is destined for the striate cortex of the occipital lobes, located at the back of the brain.

To get there, the axons of the ganglion cells exit the eye via the optic nerve and travel backwards out of the orbit into the cranial cavity. Within the skull, the two optic nerves meet at a junction called the optic chiasm, where the fibres from the nasal side of each retina swap sides, resulting in all the information from the left half of the visual field going to the right cerebral hemisphere, and vice versa. From the optic chiasm a small percentage of the visual information is relayed to a nucleus in the brainstem called the superior colliculus, which helps regulate eye movements. The remainder relays first at the lateral geniculate nucleus of the thalamus before transferring to the primary visual cortex ("V1") located in the occipital cortex.[21]

Mapping of visual data from the retina to the occipital cortex preserves a topographical relationship (retinotopic mapping), but the central area of the visual field (corresponding to foveal coverage) receives significantly greater representation within the cortex compared to the peripheral regions of the visual field, known as the cortical magnification factor.[25]

From the primary visual cortex, visual sensory data is transferred via two main neural pathways: the ventral stream and the dorsal stream. The ventral stream connects V1 to the inferior temporal cortex via other visual centres in the occipital lobe, and has been called the "What Pathway", being associated with form recognition and object representation. The dorsal stream connects V1 with the posterior parietal cortex (also via centres in the occipital lobe) and is known as the "Where Pathway" or "How Pathway". It is concerned with motion, representation of object locations, and control of the eyes and arms, especially when visual information is used to guide saccades or reaching.[26]

It is at this point in our journey from eye to brain that it starts to feel a bit like chasing one's own tail, as the "pure" visual information from the retina starts to get increasingly diluted in numerous different brain regions, and "contaminated" by data or other modulation related to other bodily activity or cognitive processes.

There is ongoing compression of visual data between the optic nerve and the visual cortex, with only 10^4 bits of data per second reaching layer 4 of V1. However, while the original retinal data seems to peter out the further we get into the brain, the number of visual-related synaptic connections progressively multiplies. By way of example, the number of synapses in the lateral geniculate nucleus of the thalamus and in layer IV of the primary visual cortex devoted to incoming visual information is less than 10% of the total number of synapses in both locations.[27]

It would certainly seem that the light entering our eyes is only one contribution to what we experience as visual reality. Given the vast amount of visual data landing on our retina from moment to moment, it seems highly counterintuitive to find that estimates of the bandwidth of conscious visual awareness are in the range of only 100 bits per second or less.[28]

Part of our brain's contribution to our visual experience is to actively suggest or impose certain interpretations onto the world around us. Take a look at Figure 2.8, showing a range of optical illusions. We can't help but see a cylinder, triangle, Loch Ness Monster, sphere, and circle. To paraphrase science writer James Burke, reality is partially constructed in the brain before it is experienced.[29]

The cumulative effect of corrective eye movements, perceptual interpolation, lateral inhibition, and a whole host of cerebral visual processing is that the human visual system is actively engaging with the world outside the body – it cannot be considered a passive entity awaiting visual information to enter in a disinterested fashion. Our anatomy and physiology confines and constrains what we are able to see but is adapted to maximise what we understand of our visual world. In doing so there is a significant amount of presupposition and guesswork.

HUMAN AND NON-HUMAN VISION

In reading this sentence, your own experience will tell you that despite everything I have just said, human vision seems to be pretty good. This reflects a combination of the visual system being able to deliver veridical images in a wide variety of lighting

FIGURE 2.8 Series of optical illusions, demonstrating illusory contours. Subjective contours are visual illusions that evoke the perception of an edge without a luminance or colour change across that edge. Image credit: Peter Hermes Furian/ Shutterstock

conditions (i.e. it really *is* pretty good), and the human brain being highly adept at "glossing over" those situations in which it struggles to do so. We might also take some comfort from a paper suggesting humans can resolve four to seven times more detail than dogs and cats, and more than a hundred times more than a mouse or a fruit fly.[30]

However, when we survey the animal kingdom it becomes clear that the human visual system cannot be considered the definitive way to see the world. The enormous variety of visual systems found in different organisms demonstrate that for any specific criteria we may wish to consider there is always an organism with superior visual ability. Conversely, animals with vision characterised as worse than our own may have other superior sensory systems such as olfaction or echolocation.[14]

Human vision is confined to a relatively narrow band of the electromagnetic spectrum which we are familiar with as visible light, the range of wavelengths being between 400 and 700 nm.[31] For a range of other organisms including birds, fish, and many arthropods, this range extends into the ultraviolet spectrum at 320–400 nm. Many flowers display striking markings only visible in ultraviolet light – imperceptible to humans but visible to the insects that pollinate these plants.

At the other end of the spectrum, infrared radiation with wavelengths between 700 and 760 nm is visible to some species of fish and butterflies.[14]

A number of animal species are also able to detect the plane of polarisation of light. Linear polarising sensitivity is used for a variety of purposes including navigation, sexual signalling and detecting water, most commonly amongst invertebrates such as arthropods and cephalopods, but is also described in vertebrates, including fish, birds, amphibians, and reptiles.[32]

The mantis shrimp has an extraordinary visual system, being able to perceive from deep UV to far-red (720 nm). Using between 12 and 16 different types of photoreceptors, it is able to modify the sensitivity of its long-wavelength colour vision to adapt to the surrounding environment by a process known as spectral tuning, a process only known in mantis shrimp. *Gonodactylus smithii*, the purple spot mantis shrimp, is unique in having dynamic polarisation vision, achieved by creating two separate streams of visual information from each eye which move independently of one another on stalks.[33]

FLICKER FUSION RATE

Flicker fusion rate, also referred to as critical fusion frequency, or the flicker fusion threshold, is a measure used by researchers in a wide variety of disciplines in both humans and in animals. It is defined as the frequency above which a flickering light source appears to be continuous to an observer. On an economic basis, its most important application is in relation to designing television and computer screens but it also has applications in studying human diseases such as Alzheimer's disease and multiple sclerosis. It provides a measure of how rapidly adaptable the visual system of the subject is.[34]

Research has shown there to be wide variation in the flicker fusion rate across different animal species. For humans, the rate is typically around 60 frames per second (with a range of 50–90 Hz), although under certain circumstances a rate of 500 Hz has been reported. Birds such as the pied flycatcher have a rate of 145 Hz, and the blowfly manages 200 Hz. The fire beetle puts the typical human performance to shame with an impressive flicker fusion rate of 400 Hz. A broad trend that has been identified is that the smaller the animal, the higher the flicker fusion rate.

It has been suggested that this variation in the "frame-refresh rate" may have a corresponding effect in relation to how different animal species experience the passage of time, with smaller animals effectively seeing events unfold in slow motion (at least when compared to our own experience).[35] This idea is explored in the animated family movie *Epic* (dir. Chris Wedge, 2013).

Although speculation on the temporal dimension of different animals' perceptual experience can run into difficulties, it certainly seems plausible that animals of different sizes and metabolic rates do gauge the passage of time at different rates. It is also interesting to note the range of variability in the flicker fusion rate amongst human subjects raising the possibility we may experience the visual passage of time a little differently from one another. Variability of the threshold within one individual under different conditions also adds credence to experiencing high stress or dangerous events in "slow motion".

The variation in flicker fusion rate in different animal species also underlines how each visual system is adapted to the ecological niche occupied by each organism. While we may be suitably impressed by the visual acuity of a bird of prey, or the comprehensive range and nuances of light detected by a mantis shrimp, trying to work out which animal has "the best" vision is a misguided concept. Each system suits the specific needs of that organism both in relation to the habitat in which it lives and the role it plays within the ecosystem. All visual systems of any organism are "good enough" to enable that species to survive, yet none can be considered "perfect" as there are always compromises.

EYE POSITION AND PUPIL SHAPE

For example, the position of eyes in relation to an animal's head varies according to the role of the organism; predators' eyes both being located frontally, prey animals having each eye located at the side of the head. Even the shape of the pupil within the eye varies according to role – ambush predators such as cats and foxes typically having a slit-like vertically aligned pupil to enable a narrow focus on potential prey directly in front of them, prey often having their pupil aligned horizontally to enable a cinemascope-style wide angle survey of the largest angle field of view. In each case there is selectivity in what can be optimally seen. The predator gains spatial resolution and superior depth perception but at the loss of a breadth of peripheral vision, and vice versa for the prey.[4]

In natural selection, it is those traits which on the balance of probability offer a survival advantage which will tend to be preserved. So, while there may be situations in which a predator loses out on the basis of such narrow, focused vision and likewise prey may founder due to taking too broad a world view, in most situations each visual system works well. A neat feature of herbivores with horizontal pupils, such as goats, deer, horses, and sheep is that as they bend their head down to eat, their eyes rotate by up to 50 degrees (known as cyclovergence), keeping the pupil aligned with the horizon.[36]

While considering predators hunting prey, I am reminded of the ecological energy pyramid diagrams from biology lessons, in which only 10% of energy is conserved between each consumer in a habitat, such that a quaternary consumer (or typical "top predator") is manifest as only 0.01% of the biomass of the plants required to keep it alive. The inefficiency always struck me as frustratingly wasteful, even though it is (very literally) an inescapable fact of life.

Likewise, when we recall that half of the light reaching the surface of the eye will be absorbed or scattered before it reaches the retina; that of 10 billion bits of information reaching the retina, only 6 million bits will be transferred via the optic nerves; and that of these perhaps only 100 bits constitute conscious perception, it would seem we are ignoring (or at least unaware of) a lot more than we are ever claiming to see, and imagining a lot more than we can claim to be observing.

In any case, our brief survey of vision in other animal species demonstrates a rich diversity in how the world can be perceived. To conclude the chapter, we will consider some variations in human visual experience.

APHANTASIA/HYPERPHANTASIA

Aphantasic individuals do not produce visual images within the "mind's eye" as the majority of the population can. For some, this occurs as the result of a pathological process in which this visualisation process is lost, having previously been present, and could be considered a disability (acquired aphantasia). However, for most aphantasic individuals (estimated to be around 1 in 50 people) this internal visualisation process has never been present (congenital aphantasia) and can be considered a normal variant of humanity's neurophysiology.[37]

Francis Galton is thought to have been the first to describe this phenomenon in the late 19th century,[38] but it has come to scientific and public attention in the past decade as a result of the work of Prof Adam Zeman, who helped coin the term aphantasia (using Aristotle's term for the mind's eye, "*phantasia*", and the prefix "*a*" to denote absence).[37] Zeman and colleagues authored a case report within the medical literature describing a man in his mid-sixties who lost the ability to generate images in his mind following an angioplasty procedure.[39] Following a summary of this case in a popular US science magazine public awareness has steadily intensified, with thousands of people recognising themselves as aphantasic and getting in touch with Zeman's team.

In addition to those who cannot generate internal images, a large number of people also reported the related ("mirror image") phenomenon of hyperphantasia, in which images generated in the mind's eye are particularly vivid and in some circumstances indistinguishable from externally perceived images. Hyperphantasia is characterised by "strong face recognition ability and autobiographical memory; a tendency to work in creative professions; an increased chance of synaesthesia; a liability to be waylaid by imagined worlds; a risk, at times, of confusing real with imagined events".[40] This contrasts with traits associated with aphantasia, including prosopagnosia (difficulty recognising faces), a reduced ability to recollect memorable events, but with an increased focus on the here and now, and strength in abstract and mathematical thought processes.[40]

People with both aphantasia and hyperphantasia have reported close relatives to also have similar respective extreme visualisation experiences at a higher rate than expected by chance suggesting a likely (though as yet unproven) genetic component.[40] However, there is considerable variability in the degree to which the described features are manifested in any one individual, so any inherited influence is likely to be polygenic in nature. Some cases of acquired aphantasia relate to psychological trauma so environmental factors are also likely to be important.

Collectively at any rate, it seems there is a spectrum of mind's eye visualisation with aphantasia at one end and hyperphantasia at the other. Of note, around half of aphantasic individuals do not have any visual dreams, whilst the remainder does. An over simplified characterisation of this spectrum would place analytical, unimaginative people focused on the here-and-now at the aphantasic end and daydreaming, creative types with some difficulties discriminating fact from fantasy at the hyperphantasic end. However, this is problematic in various ways not least because personality traits are not determined by visual perception processes (or at least not in isolation), and such traits tend not to be exclusive to any one grouping.

Characterising aphantasic individuals as visually unimaginative certainly turns out to be wide of the mark. Numerous artists, sculptors, architects, and authors have been identified in the past few years. The *Extreme Imagination* exhibition hosted

FIGURE 2.9   Elina Cerla. *Aphantasia*. Charcoal on prepared paper. 60 × 80 cm. (Photo: Justin Webb). Courtesy of the artist.

in Glasgow and Exeter in 2019 showcased the work of 16 aphantasic artists and writers alongside six hyperphantasic artists. One of these works, *Aphantasia*, by Elina Cerla is reproduced in Figure 2.9. Cerla is aphantasic with no "object imagery" but describes being able to imagine spatial reconstructions of objects and compositions.[41] Perhaps the most famous example of an aphantasic artist is Glen Keane, the animation artist who drew the lead character Ariel, in Disney's *The Little Mermaid* (dirs. Ron Clements, John Musker, 1989).

   In the preparation of this text, I contacted Prof Zeman to enquire if he was aware of any aphantasic radiologists, at which time he was not (though did point out his database of aphantasic individuals is growing rapidly). Shortly prior to my publisher's deadline, I found out that one of my departmental colleagues is, in fact, aphantasic, a revelation that poses all sorts of questions concerning how images viewed "in real time" are compared to those from the past. My own experience of performing interpretation of radiological images would suggest a need to draw upon a stock of mental images constructed in the mind's eye to draw comparisons with normality or particular disease appearances (discussed in Chapter 6). However, the emerging complexity of how visual perception and language are intertwined, exemplified by my colleague's experience, demonstrate that such comparisons do not require conscious "display" of previous images in the mind.

DEFICITS OF HIGHER VISUAL PROCESSING

A wide range of deficits or variations in human vision has offered insights into the mechanisms of higher visual processing, although often posing more questions than answers. These include:

- Simultagnosia, in which individuals can see only one object at a time and sometimes only portions of objects, unaware of focusing on just one part of a larger form
- Astereopsis – the inability to perceive depth in three-dimensional space
- Prosopagnosia – the inability to recognise faces
- Acquired colour blindness, where loss of function within the brain (rather than an issue with the photoreceptor cells) causes an inability to perceive colour

CORTICAL BLINDNESS

In contrast to the bullet-pointed deficits above, in which particular attributes of vision are affected, cortical blindness describes a total (or near-total) inability to see as a result of damage to visual processing areas in the brain, most commonly as a result of an insult to the occipital cortex (or cortices). However, within this broad diagnostic category, there are some very interesting variations.

Anton syndrome (also known as Anton-Babinski syndrome) describes patients with neurological visual impairment/disturbance resulting from abnormality or damage in the brain rather than due to eye abnormalities, but in whom there is denial of any visual loss ("visual anosognosia"), often associated with confabulation (the invention of fictional memories of events and experiences). The first case report of Anton syndrome dates to 63 CE, authored by Seneca, a Roman philosopher, politician, and adviser to Emperor Nero, who later had him executed. He describes the case of Harpaste, his wife's slave who acutely became blind.

> The story sounds incredible, but I assure you that it is true: she does not know that she is blind. She keeps asking her attendant to change her quarters; she says that her apartments are "too dark". [42]

This syndrome is rare but is now well established within modern medical literature.

A contrasting phenomenon is that of blindsight, whereby patients who are clinically blind due to damage to the primary visual cortex, are able to detect, localise, and even discriminate visual stimuli that they deny seeing. This demonstrates that while a functioning primary visual cortex (known as V1) is required for conscious visual perception, our visual system in its entirety is still capable of processing and, in some circumstances, acting upon visual information when the capacity for conscious awareness is no longer present.[43] An implication of this research is that even those of us with fully functioning visual cortices may be, at times, acting on visual information that we have not consciously been aware of.

So, in aphantasia/hyperphantasia we have encountered individuals who are unable to generate internal visual images on the one hand and those who may struggle to separate imagined visual images from reality on the other. In cortical blindness, there are some individuals who claim to be able to see despite incontrovertible evidence to the contrary, and others who are unable to consciously see anything, yet can perform tasks which require visual processing.

A variant of cortical blindness is Statokinetic dissociation (also called Riddoch phenomenon or Riddoch syndrome) – the ability to perceive visual motion consciously in a blind visual field.[44] The combination of patients with Riddoch syndrome with those suffering from akinetopsia – the inability to perceive motion, despite being able to see static objects clearly, provides evidence for the processing of motion as a discrete visual characteristic at a site (or sites) in the brain distant from the primary visual cortex. Likewise, loss of specific characteristics of vision such as depth perception and colour is consistent with a modular system of visual processing whereby different areas of the brain make sense of different imaging attributes, and the job of the primary visual cortex is to bundle up all these different components to produce a coherent visual consciousness.

SUMMARY

While human vision offers an extraordinary sensory experience, being both fluid and immersive we have seen that it is limited in its scope, range of sensitivity and based on a high degree of selectivity. The rich and detailed appearances we enjoy far outperform those of a low-grade digital camera – the anticipated image resolution based on the output of retinal photoreceptors alone, indicating substantial "postprocessing" performed by the brain. In making sense of our visual environment we are also highly dependent on presupposition and joining up dots, imposing "best-fit" hypotheses on all that we survey.

REFERENCES

1. S. Standring (ed), *Gray's Anatomy*. 40th edition, Churchill Livingstone, London, 2008, Chapter 40 – The eye. pp 675–697.
2. M.S. Sridhar, Anatomy of cornea and ocular surface, *Indian J Ophthalmol.*, 66(2), 2018, 190–194. doi:10.4103/ijo.IJO_646_17
3. M. Goel, R.G. Picciani, R.K. Lee and S.K. Bhattacharya, Aqueous humor dynamics: A review, *Open Ophthalmol J.* 4, 2010, 52–59. Published 2010 Sep 3. doi:10.2174/1874364101004010052
4. M.S. Banks, W.W. Sprague, J. Schmoll, J.A.Q. Parnell and D.G. Love, Why do animal eyes have pupils of different shapes?, *Sci. Adv.* 1(7), 07 Aug 2015, e1500391. doi:10.1126/sciadv.1500391
5. N.K. Tram and E.K. Swindle-Reilly, Rheological properties and age-related changes of the human vitreous humor front. *Bioeng. Biotechnol.* 18(6), 2018, 199. doi:10.3389/fbioe.2018.00199
6. J.G. Goldman, 18th January 2016, www.bbc.com/future/article/20160113-why-do-you-get-eye-floaters

7. R.L. Gregory, *Eye and Brain: The Psychology of Seeing*. 5th edition, Princeton University Press, Princeton, 1997.
8. D. Mustafi, A.H. Engel and K. Palczewski, Structure of cone photoreceptors, *Prog Retin Eye Res*. 28(4), 2009, 289–302. doi:10.1016/j.preteyeres.2009.05.003
9. S.E. Palmer, *Vision Science: Photons to Phenomenology*, The MIT Press, Cambridge, 1999.
10. Google Search, How far away is the andromeda galaxy in mileswww.google.com/search?q=how+far+away+is+the+andromeda+galaxy+in+miles&rlz=1CACCCC_enG B868&oq=how+far&aqs=chrome.1.69i57j69i59j0l6.4820j1j7&sourceid=chrome&ie= UTF-8, last accessed 21/07/21.
11. E. Ribak, Light propagation explains our inverted retina, *SPIE Newsroom 01*, 01, 2010. doi:10.1117/2.1201009.003189
12. S.E. Palmer, *Vision Science: Photons to Phenomenology*, The MIT Press, 1999 [loc 1673].
13. M.Mazurak and J. Kusa, Jan Evangelista Purkinje: A passion for discovery, *Tex Heart Inst J*. 45(1), 2018, 23–26. Published 2018 Feb 1. doi:10.14503/THIJ-17-6351
14. M.F. Land and D.-E. Nilsson, *Animal Eyes*, Oxford University Press, Oxford, 2012, p 18.
15. T.D. Lamb, S.P. Collin and E.N. Pugh Jr, Evolution of the vertebrate eye: Opsins, photoreceptors, retina and eye cup, *Nat Rev Neurosci*. 8(12), 2007, 960–976. doi:10.1038/nrn2283
16. J.M. Provis, A.M. Dubis, T. Maddess and J. Carroll, Adaptation of the central retina for high acuity vision: Cones, the fovea and the avascular zone, *Prog Retin Eye Res*. 35, 2013, 63–81. doi:10.1016/j.preteyeres.2013.01.005
17. D.E. Sabih, A. Sabih, Q. Sabih and A.N. Khan, Image perception and interpretation of abnormalities; can we believe our eyes? Can we do something about it? *Insights Imaging* 2(1), 2011, 47–55. doi:10.1007/s13244-010-0048-1
18. H.K. Hartline, H.G. Wagner and F. Ratliff, Inhibition in the eye of Limulus. *J Gen Physiol*. 39(5), 1956, 651–673. doi:10.1085/jgp.39.5.651
19. P. Pojman, Ernst Mach, *The Stanford Encyclopedia of Philosophy*, Spring 2019 edition, E.N. Zalta (ed.), https://plato.stanford.edu/archives/spr2019/entries/ernst-mach/.
20. J.D. Kropotov, Chapter 15 – Methods: Neuronal networks and event-related potentials, in *Quantitative EEG, Event-Related Potentials and Neurotherapy*, 2009, pp 325–365, doi:10.1016/B978-0-12-374512-5.00015-3
21. V. Bruce, P.R. Green and M. A. Georgeson, *Visual Perception: Physiology, Psychology and Ecology*. 4th edition, Psychology Press, Taylor and Francis Group, 2003 [loc 822].
22. S. Standring (ed), *Gray's Anatomy*. 40th edition, Churchill Livingstone, 2008, Chapter 39 – The orbit and accessory visual apparatus. pp 655 –674
23. D. Purves, G.J. Augustine and D. Fitzpatrick, et al., (eds), *Neuroscience*. 2nd edition. Sinauer Associates, Sunderland, MA, 2001. Types of eye movements and their functions. www.ncbi.nlm.nih.gov/books/NBK10991/
24. S. Zeki, Bridget Riley and the art of the brain, in *About Bridget Riley: Selected Writings 1999–2016*, D. Globus and K. Schubert (eds), Ridinghouse, 2017, p 402.
25. S.E. Palmer, *Vision Science: Photons to Phenomenology*, The MIT Press, 1999 [loc 1779].
26. M.A. Goodale and A.D. Milner, Separate visual pathways for perception and action, *Trends Neurosci*. 15(1), 1992, 20–25. doi:10.1016/0166-2236(92)90344-8. PMID 1374953.
27. M. E. Raichle, The restless brain: How intrinsic activity organizes brain function, Published: 19 May 2015, doi:10.1098/rstb.2014.0172 https://royalsocietypublishing.org/doi/10.1098/rstb.2014.0172

28. T. Nørretranders, *The User Illusion: Cutting Consciousness Down to Size*, Sydenham, Viking, NY, 1998.

29. J. Burke, *The Day the Universe Changed*, Little, Brown and Company, Boston, Toronto, 1985, p 308.

30. E.M. Caves, N.C. Brandley and S. Johnsen, Visual acuity and the evolution of signals, Published: March 30, 2018, doi:10.1016/j.tree.2018.03.001

31. V. Bruce, P.R. Green and M.A. Georgeson, *Visual Perception: Physiology, Psychology and Ecology*. 4th edition, Psychology Press, Taylor and Francis Group, Abingdon, 2003 [loc 317].

32. J. Marshall and T.W. Cronin, Polarisation vision, *Curr. Biol.* 21(3), 8 February 2011, R101–R105. doi:10.1016/j.cub.2010.12.012

33. Wikipedia, Mantis Shrimp, https://en.wikipedia.org/wiki/Mantis_shrimp, last accessed 21/07/21

34. G. Ganesh, S. Mahalingam, G. Annamalai and U. Damodharan, Seeing is believing: A demonstration of critical fusion frequency and its multidimensional nature, *Adv Physiol Educ.* 41(2), 2017, 315–319.

35. K. Healy, L. McNally, G.D. Ruxtond, N. Cooper and A.L. Jackson, Metabolic rate and body size are linked with perception of temporal information, *Anim. Behav.* 86(4), October 2013, 685–696.

36. G. Love, Revealed: Why animals' pupils come in different shapes and sizes, Durham University, 10/08/15, www.dur.ac.uk/news/allnews/thoughtleadership/?itemno=25 392#:~:text=So%2C%20vertically%20elongated%20pupils%20help,prey%20ani mals%20avoid%20their%20predators.&text=We%20found%20that%20eyes%20of,p upil%20aligned%20with%20the%20ground. Last accessed 21/07/21

37. A. Zeman, M. Dewar and S. Della Sala, Lives without imagery – Congenital aphantasia, *Cortex* 73, 2015, 378–380. doi:10.1016/J.cortex.2015.05.019

38. F. Galton, Statistics of mental imagery, *Mind* 5, 1880, 301–318.

39. A. Z. Zeman, S. Della Sala, L. A. Torrens, V. E. Gountouna, D. J. McGonigle and R. H. Logie, Loss of imagery phenomenology with intact visuo-spatial task performance: A case of 'blind imagination', *Neuropsychologia* 48(1), 2010, 145–155. doi:10.1016/j.neuropsychologia.2009.08.024

40. A. Zeman, Phantasia: The (re)discovery and exploration of imagery extremes, pp 6–13, Catalogue: Extreme Imagination – inside the mind's eye, First published in Great Britain in 2018 by: The Eye's Mind University of Exeter College of Medicine and Health, Exeter, EX1 2LU.

41. E. Cerla. Commentary p6 (reversed face) Catalogue: Extreme Imagination – inside the mind's eye.

42. C. André MD, PhD, Seneca and the first description of anton syndrome, *J Neuro-Ophthalmol.* 38(4), December 2018, 511–513. doi:10.1097/WNO.0000000000000682

43. Ajina S. and Bridge H., Blindsight and unconscious vision: What they teach us about the human visual system, *Neuroscientist.* 23(5), 2016, 529–541. Published 2016 Oct 23. doi:10.1177/1073858416673817

44. R. Hayashi, S. Yamaguchi, T. Narimatsu, H. Miyata, Y. Katsumata and M. Mimura, Statokinetic Dissociation (Riddoch Phenomenon) in a patient with homonymous Hemianopsia as the first sign of posterior cortical atrophy, *Case Rep Neurol.* 9(3), 2017, 256–260. Published 2017 Nov 10. doi:10.1159/000481304

3 Art, Artefact, and Artifice

In conversation with art historian Sir John Leighton to celebrate a major retrospective exhibition of her work in 2019, abstract artist Bridget Riley recounted a pivotal episode early in her career. At a low moment of immense frustration and bitter disappointment she attempted to convey her feelings through a work composed entirely of formless black brushstrokes.

> I painted it in fury, a black canvas with what amounted to a sort of very expressionist handling of black paint. I came down the next morning and looked at it, and I realised that it was quite inexpressive of what I felt, it was just a lot of black paint. It said nothing. It conveyed nothing. [1]

Riley's reflection on the failure of this work (which she later destroyed) led her to a new approach in which an emphasis on form, opposition, and contrast were key. Soon after this episode she passed on the hands-on production of her work to studio assistants demonstrating a clear distinction between a creative vision and the craft required to bring it into being.[2]

This journey from an entirely expressionistic approach in the black canvas work, where the movement of the brushstrokes in an otherwise formless composition was intended to convey meaning, to crisp "pure" images, untouched by her own hand serves to introduce us to two different approaches to how art can be executed.

ÉCRITURE

The first approach allows, or perhaps actively encourages, the mark of the artist to be visible within the finished work of art. This might include brushstrokes which are clearly identifiable as such, or directional linear strokes of pen, pencil, or pastel, including visible cross-hatching. Paint being applied so thickly that it rises up from the canvas, usually associated with identifiable brushstrokes or indentations of a palette knife, has earned its own descriptive term *impasto*.

Art historians have drawn a distinction between visible marks of a human hand having been involved in the production of a piece and the unique stylistic signature contained within those marks which help identify a work as having been produced by a specific individual, using *écriture* (the French word for handwriting) for the former, and *touche* for the latter.

The work of Vincent Van Gogh can be used to illustrate these terms. The Dutch post-impressionist is perhaps the most celebrated artist in relation to *écriture* – his instantly recognisable and much copied style of discrete, bold brushstrokes with daubs of impasto oil paint lifting in relief from the canvas achieves a dynamic and distinctive feel to each work.

DOI: 10.1201/9780367855567-3

He described his technique in a letter to his brother:

I follow no system in painting … I hammer away at the canvas with irregular strokes and let it all stay as it is – here and there impasto, raw canvas in places, unfinished corners. Brutalities![3]

This is brought to vivid effect in the movie *Loving Vincent* (dirs. Dorota Kobiela and Hugh Welchman, 2017), in which 65,000 frames of film were hand painted in this very particular style, such that the entire story appears to have been painted in oil rather than filmed, with daubs of colour melting into one another as each scene unfolds.[4] The irregular strokes and impasto elevation from the canvas can be characterised as the *écriture* (techniques by no means unique to Van Gogh), whereas a combination of the distribution and pattern of the strokes together with the colour allocation produced his distinctive *touche*, such that the two-dimensional images in *Loving Vincent* are able to effectively replicate his style.

Figure 3.1, a painting of an old pair of shoes (1888), illustrates his style well. The shading to the right of the shoes is particularly distinctive, achieved by numerous bold linear brushstrokes in series and in parallel with one another. These are similar in feel to "action lines" used in comics and cartoons to denote movement or

FIGURE 3.1 Oil painting of an old pair of shoes (1888), by Vincent Van Gogh. Shoes, by Vincent Van Gogh, 1888, Dutch Post-Impressionist, oil on canvas. Image credit: Everett Collection/ Shutterstock

sometimes an emotion such as shock or surprise and do add dynamism to a seemingly mundane choice of subject matter.

Van Gogh is known to have been influenced by Japanese art and is said to have coined the term *Japonaiserie* to describe the broader influence of this work on contemporary European artists.[5, 6] Figure 3.2 (Temple procession to Torinomachi in the Rice Fields of Asakusa, by Hiroshige and Utagawa, 1857) is a print from a woodcut embodying such influences. Although the finished print resulting from a woodcut would not fit into the *impasto* group, the original mark of the artist is nevertheless visible in the deliberative nature of each line visible in the final image, a feature we will return to. Another example of Japanese art is shown in Figure 3.3, a watercolour painting of a bamboo plant in which almost every mark on the image is identifiable as an individual brushstroke. Likewise, traditional Chinese art placed an emphasis on specific characteristic brushstrokes, with the perceived quality of work often judged in large part on the mastery of these particular brushstrokes. Indeed, within this Chinese tradition, an attempt to conceal or diminish the visibility of such brushstrokes was frowned upon as a dishonest technique, and deception of this sort was viewed as ultimately detracting from the truth of the image.[7]

Other artistic movements in which the mark of the artist – in one form or another – is paramount include primitivism, expressionism, abstract expressionism, and neo-expressionism.

PURE ART

The second approach takes a different view. In this approach, any visible mark of the artist diminishes the value (and truth) of the work.

Compare the shovelling-paint-on-with-a-trowel approach used by Van Gogh to the view of his 19th-century contemporary James Abbott McNeill Whistler:

"Paint should not be applied thick. It should be like a breath on the surface of a pane of glass".[8]

Earlier in the 19th century, John Constable had been advised by Joseph Farington (a senior member of the Royal Academy of Arts) that his work would be better received if his brushstrokes were less evident, reflecting a broader fashion of the late 18th and early 19th century that David Hockney characterised as "the elimination of the brushstroke".[9]

While the brushstroke was to return in a big way by the end of the 19th century, the urge to minimise or eliminate the visible mark of the artist continued in 20th-century movements such as conceptualism and minimalism. As already mentioned, Bridget Riley has pursued works in which no trace of manual manipulation or craftsmanship is visible in the final pieces, the physical incarnation of her vision being executed by her assistants ensuring the mark of the artist, in the traditional sense, is absent.

Eliminating all trace of the artist in relation to a finished work had earlier been taken to a logical conclusion of sorts by Marcel Duchamp with his "readymades"

FIGURE 3.2   Temple procession to Torinomachi in the Rice Fields of Asakusa, by Hiroshige 1st and Utagawa, 1857, Japanese woodcut. A white cat looking outdoors from the window, with Mount Fuji in the distance. Image credit: Everett Collection/ Shutterstock

FIGURE 3.3 A watercolour painting of a bamboo plant in which almost every mark on the image is identifiable as an individual brushstroke. The painting is from an album of 18th century Japanese watercolours of plants and landscapes. Credit: Wellcome Collection. Attribution 4.0 International (CC BY 4.0).

in which pre-existing objects (for which he had no input in relation to their design or manufacture) were displayed without alteration or modification (with the exception of adding the artist's signature and a title). The act of choosing the object to be displayed – most notoriously a urinal in *The Fountain* – was the artistic endeavour. An explanation of the readymade was delivered in an anonymous leading article in the May 1917 edition of *The Blind Man* an avant-garde magazine run by Duchamp and two friends:

> Whether Mr Mutt with his own hands made the fountain or not has no importance. He CHOSE it. He took an ordinary article of life, and placed it so that its useful significance disappeared under the new title and point of view – created a new thought for that object.[10]

Many have questioned the artistic merit of this approach, and the "readymade" was successfully satirised by teenagers Kevin Nguyen and T.J. Khayatan, in 2016.

Nguyen placed his sunglasses on the floor of the San Francisco Museum of Modern Art and the pair then stood back admiring their prank exhibit, taking photographs and in doing so managing to convince other visitors that the sunglasses were in fact a work of art.[11] It is instructive to note that Nguyen's actions do in fact meet Duchamp's criteria for creating art (demonstrated by replacing "Mr Mutt" with "Mr Nguyen", and "fountain" with "sunglasses" in the quote above), so perhaps the visitors who gathered round the sunglasses were more discerning than the reports of this episode would have us believe. While readymade and conceptual art will doubtless continue to divide opinion, the purity of the medium – untouched by the artist – does have a compelling internal logic.

ART AND ARTEFACT

Now let's return to the arena of medical imaging and think about which of the two approaches we have considered is best suited to providing the most accurate representation of the body, and the anatomical and pathological reality that the radiologist is trying to make sense of.

If you can forget urinals and spectacles on the floor of art galleries, it is intuitive to think that eliminating extraneous features from the image would be the best approach. Instead of talking about the mark of the artist or écriture, we will talk about imaging artefact. This is effectively the "mark of the scanner" – visible features on the scan image that in theory should not be there. These can be found in all the imaging modalities currently used in clinical practice. In the same way that the Royal Academy would argue brushstrokes should not be visible, or that Whistler would advocate the very thinnest layer of paint, rather than it bulging up from the canvas, under most circumstances radiographers and radiologists are seeking to produce and interpret artefact-free images. The anatomical truth should not be tarnished by imperfections of the scanning technology.

There are certainly a large number of different imaging artefacts that do frequently distort anatomy, and in some situations these can both mimic and mask pathology. In paediatric imaging, motion artefact is amongst the most frequently encountered, illustrated in Figures 3.4–3.7. Figures 3.4 and 3.5 are CT topograms – an X-ray-based image used to plan the formal CT examination. In Figure 3.4 the upper limbs are distorted by the child having moved, and in Figure 3.5 the skull is likewise seemingly stretched by movement during the scan. Figure 3.6 is a CT scan of a child's head, rendered incomprehensible by motion artefact and Figure 3.7 is an MRI of the head similarly fragmented and overlapped. Although you might argue that in these cases it is the "mark of the patient" rather than the scanner itself that is to blame for the artefact, the higher the acquisition time (determined by the nature of the scan), the higher the likelihood of motion artefact occurring. While the combined skills of radiographers, radiology department assistants, and play therapists usually enable most children to tolerate short duration imaging examinations, MRI scans requiring a subject to lie still for minutes at a time often necessitate a general anaesthetic in younger children.

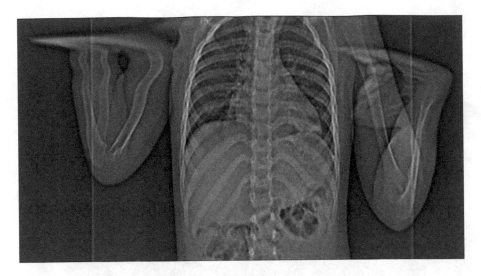

FIGURE 3.4 CT topogram of a child's chest, in which the upper limbs are distorted by motion artefact. Image provided by the author.

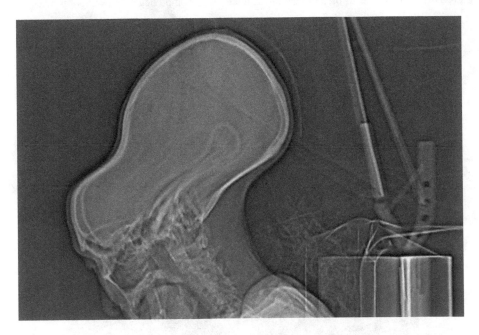

FIGURE 3.5 CT topogram of a child's head, in limbs which the skull is seemingly stretched by movement during the scan. Image provided by the author.

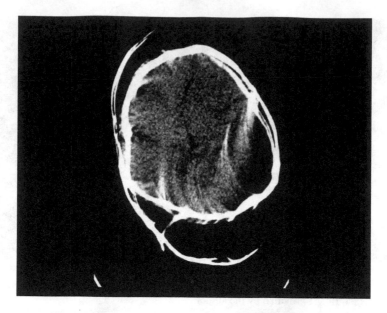

FIGURE 3.6 CT scan of a child's head, rendered incomprehensible by motion artefact. Image provided by the author.

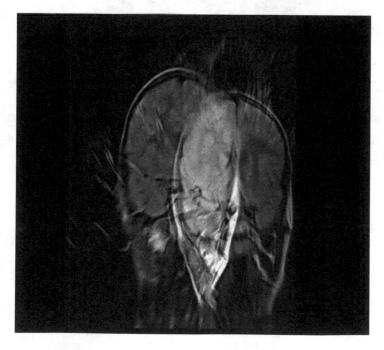

FIGURE 3.7 MRI of a child's head, fragmented and overlapped due to motion artefact. Image provided by the author.

However, while an artefact can certainly be a nuisance with the potential to mislead, it can also provide extremely useful clinical information. In ultrasound, for example, the related artefacts of acoustic shadows and posterior acoustic enhancement can help radiologists and sonographers make sense of what they are seeing.

Acoustic shadows are cast by dense matter, including bones and calcific structures – the majority of sound waves emitted by the transducer are reflected at the interface, such that the area beyond cannot be visualised. While this is problematic when scanning through areas with overlying bones (such as scanning parts of the upper abdomen covered by ribs) this artefact can be helpful in identifying calculi in various locations, such as gallstones, renal stones, and appendicoliths. The characteristic shadows cast by foci of calcification can also provide useful information in diagnosing certain tumours. Figure 3.8 shows a gallstone located within the common bile duct casting a distinctive acoustic shadow.

Posterior acoustic enhancement is effectively the reverse phenomenon of acoustic shadowing. This occurs when a structure containing fluid is located within the beam of the ultrasound transducer. Within areas of fluid almost all the soundwaves are transmitted, with minimal reflection. This means more soundwaves reach the tissues beyond the area of fluid than would usually be the case. More soundwaves get reflected back from this corridor of tissue, producing a characteristic bright "negative" shadow behind the area of fluid, such as can be seen deep to the cyst shown in Figure 3.9. The presence of posterior acoustic enhancement can therefore be a useful imaging feature when needing to discriminate between fluid or low reflectivity soft tissue structures.[12]

In MRI a phenomenon known as susceptibility can produce artefactual appearances, a dramatic example of which we will see shortly. The artefact is caused by small variations or inhomogeneities in the magnetic field and for the first 30 years or so of clinical use of MRI this was actively avoided with the majority of sequences being devised specifically to eliminate this problem. However, in the last 10–15 years a sequence base specifically on this principle has become established in clinical practice. Susceptibility-Weighted Imaging (or SWI) is particularly sensitive for small areas of micro-haemorrhage, and has proven to be useful in a variety of settings, including characterisation of vascular lesions and assessing the extent of traumatic brain injuries.[13] Figure 3.10 shows an example – the distinctive black blobs distributed through the brain represent areas of abnormal blood vessels called cavernomas, many of which are difficult to identify on the other MRI sequences.

For each of these cases, there is ambiguity between what constitutes the "true" image versus the artefactual appearance, and the radiologist is perhaps more concerned with what is "useful" than what is "truthful". More broadly, in an age of increasingly fabricated, artificial images, features which were once actively avoided as a corruption of the "true" image are increasingly utilised as a ploy to make such fabricated images seem more convincing, realistic, or stylish. Let us consider a few examples.

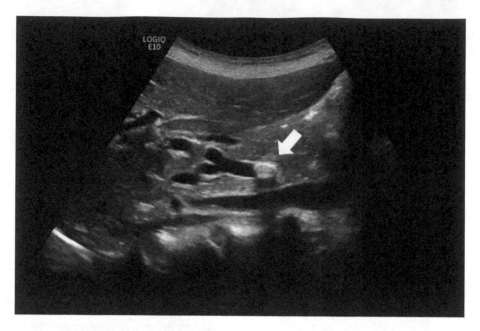

FIGURE 3.8 Ultrasound image showing a gallstone located within the common bile duct (indicated with arrow) casting a distinctive acoustic shadow. Image courtesy of Dr S Choi.

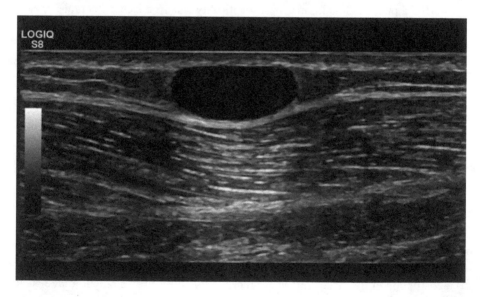

FIGURE 3.9 Ultrasound image showing a cyst located within the subcutaneous tissue. Deep in the cyst the tissue appears brighter than on either side, an imaging artefact known as "posterior acoustic enhancement". Image provided by the author.

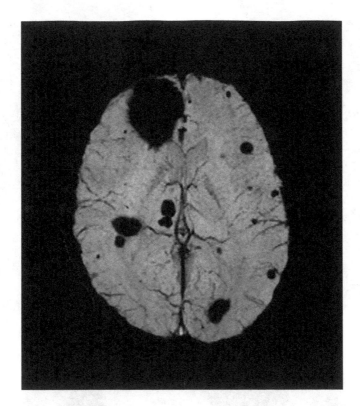

FIGURE 3.10 MRI scan of the brain acquired using Susceptibility Weighted Imaging (SWI); the distinctive black blobs distributed through the brain represent areas of abnormal blood vessels called cavernomas, many of which are difficult to identify on the other MRI sequences. Image courtesy of Dr A. Quigley.

LENS FLARE

Lens flare is an imaging artefact found in photography and cinematography which occurs when strong directional light (most often low-lying sun) reflects internally on lens elements within the camera. Light may bounce back and forth a number of times before finally reaching the film or digital sensor, producing a characteristic halo-like glare, polygonal rings, or linear streaking.[14] Although the mechanism by which this is produced is very different to streak artefact found in CT scans (encountered when a very dense material such as metal completely blocks the X-ray beam), the respective appearances share a certain resemblance. For example, compare Figure 3.11, a stock image of photographic lens flare to Figure 3.12 in which dramatic streak artefact is seen in a CT scan of a patient's head who had previously had an aneurysm coiled.

Until the late 1960s it was a cinematic convention to eliminate lens flare as an unwanted intrusion, a nuisance destined for the cutting room floor, and cinematographers would go to great lengths to avoid it. In *Cool Hand Luke* (dir. Stuart Rosenberg, 1967) cinematographer Conrad Hall purposefully broke the convention, utilising lens

FIGURE 3.11   Photographic lens flare. Image credit: donatas1205 /Shutterstock.

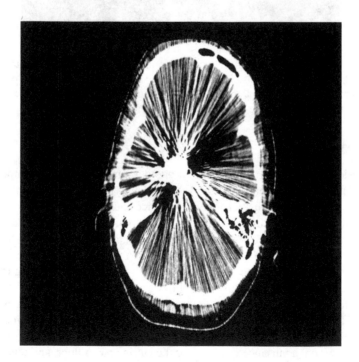

FIGURE 3.12   CT scan of a patient's head who previously had an aneurysm coiled, in which dramatic streak artefact is visible. Image provided by the author.

flare throughout the movie, particularly effective during scenes of a chain gang slaving in blazing heat, the radiant glare emphasising the intensity of the midday sun.[15] Iconic movies including *Planet of the Apes* (dir. Franklin J. Schaffner, Cinematographer Leon Shamroy, 1968) and *Easy Rider* (dir. Dennis Hopper, Cinematographer Harrison Arnold, 1969)[16] followed suit and before long what was previously an unwanted artefact became a familiar sight in mainstream commercial movies.

In recent years the effect has been heavily utilised by Hollywood giants Michael Bay and J.J. Abrams, including artificially added CGI lens flare in several of their movies. Abrams encountered some backlash to the excessive use of lens flare in his films, to the point that his wife even suggested he eased off using it.[17] Indeed, having deliberately engineered a number of shots to produce dramatic lens flare during the primary shooting of *Star Trek: Into Darkness* (Cinematographer Dan Mindel, 2013), Abrams later instructed the special effects wizards of *Industrial Light and Magic* to erase or tone down these appearances for the final cut.[18]

OFF THE GRID

While lens flare typically relates to strong directional light being shone into the camera, directors have frequently utilised lighting schemes in essentially the reverse direction, using powerful light sources behind or adjacent to the camera to produce striking contrasts and cast sharp shadows. In an interesting paper examining thematic parallels between the cinematic genre of Film Noir and radiography, Hugh S. Manon suggests:

> The hallmark of film noir lighting design … is its emphasis on the semi-permeability of spatial divides – the on-screen evocation of light passing through an aperture, or series of apertures, figured in precisely the optical / photographic sense. Whether the shadow-casting obstruction is comprised of slatted blinds, lathed stair balusters, or a window with a private investigator's name painted on it, the striking depiction of light having passed through to "X" is wholly consistent with the aura of unsuspicion I view as both noir's narrative core and its key generic difference.[19]

Furthermore,

> noir lighting techniques do more than establish a mood; they idiomatically restate, in shadow and light, the truth-exposing image that results when certain types of radiation are shot through the human body.[20]

Fans of the noir genre, in which shady characters are often to be found peering through the gaps of a venetian blind or similar (perhaps a little like the cat in Figure 3.2), may be interested in the concept of the radiographic anti-scatter grid, a device used to improve the image quality of radiographs. The device consists of very thin linear strips of radio-opaque metal aligned in parallel, evenly spaced with very narrow intervals between the strips. This is placed between the patient and the X-ray plate, and serves to block scattered X-rays (those which have diverged from their original path after interaction with atoms of the body). The grid will also block a fraction of X-rays which have not been diverted, and which, if not intercepted, could have contributed useful information to the resultant image of the radiograph. However, the proportion of scattered X-rays that are blocked by the grid is higher than that of the unscattered, so the net effect is that there is an overall reduction in image noise. Counterintuitively, the use of the grid can, in specific circumstances, allow a better quality image to be obtained using a smaller dose of radiation – despite the introduction of a device which serves to block a proportion of the light from which the image is constructed.

The similarity of a noir character surreptitiously observing goings-on through a venetian blind to a radiologist viewing an individual's body through a grid may feel a bit of a stretch, but the comparison does serve to highlight some of the philosophical dimensions of radiographic practice. The ambiguity of what constitutes a "true" image in the setting of an anti-scatter grid resonates with the tension in noir movies that Manon identifies:

> Trading back and forth across the boundaries of criminal artifice, noir confronts viewers with the impossibility of clearly dividing the world into "places of innocence" and "places of deceit".[21]

The use of an anti-scatter grid usually results in faint parallel lines being visible on the radiograph, as shown in a selected portion of a pelvic radiograph in Figure 3.13. The lines are clearly artefactual but are inseparable from the image. A duplicate radiograph acquired with the same exposure factors but without the grid in place would – although free of parallel lines – be of reduced image quality. The lines simultaneously degrade and enhance the image.

We have already seen some dramatic examples of motion artefact degrading medical images. Take a look at Figure 3.14 which shows a CT scan of a child's chest reconstructed in coronal section. Looking closely reveals that the image is composed of numerous transverse "slices" stacked up on top of one another. The movement of the child during the scan manifests itself as the slices getting out of position, similar to the bricks in a game of Jenga getting dislodged a little as successive bricks are removed. Of a similar flavour, consider Figure 3.15. Every once in a while, the

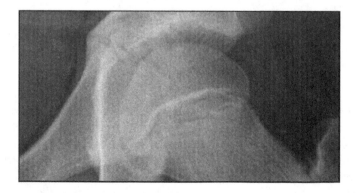

FIGURE 3.13   Cropped portion of a pelvic radiograph. The use of an anti-scatter grid has caused the faint parallel lines. Image provided by the author.

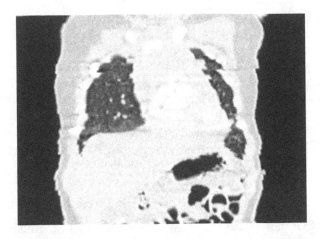

FIGURE 3.14   A CT scan of a child's chest reconstructed in coronal section. The image is composed of numerous transverse "slices" stacked up on top of one another. The movement of the child during the scan manifests itself as the slices getting out of position, similar to the bricks in a game of Jenga getting dislodged a little as successive bricks are removed. Image provided by the author.

FIGURE 3.15   Screen-grab image showing a coronal CT reconstruction of the chest loading up. At locations where the images have not yet loaded, horizontal black lines traverse the image, mimicking the appearance of a barcode. Image provided by the author.

computers at work take some time to load up reconstructed CT images, with the image appearing almost one slice at a time. This phenomenon has been captured as a screen-grab showing a CT of the chest loading up. The striking black parallel bands where slices have yet to be loaded by the computer did remind me of noir characters looking through gaps in the venetian blinds, and of the chiaroscuro lighting schemes characteristic of these movies. In a nod to such noir movies, slatted blind shadows were recently used in multiple scenes in David Fincher's *Mank* (2020).

Even if these images do not make you think of noir movies, these artefacts of image construction and their subsequent display do nevertheless illustrate the malleable nature of how anatomical reality is represented.

RAINDROPS KEEP FALLING ON MY NON-EXISTENT CAMERA

In numerous CGI movies and videogames there are a host of artificial artefacts to mimic a camera-like appearance, including raindrops on the camera screen during the replays in the FIFA football game, mud splatters in driving games, and imaginary debris obscuring the view in climactic battle sequences in action movies. The BBC series about extinct prehistoric creatures *Walking with Beasts* (2001) made playful use of such tricks, including a paraceratherium knocking over the (non-existent) camera; our ancestors, Australopithecus, cracking the camera screen whilst throwing stones at a predatory big cat; and a young mammoth covering the screen with mud squirted out of its trunk. More recently, the Pixar movie *Inside Out* (dirs. Pete Docter and Ronnie Del Carmen, 2015) replicates the jerky, unstable movement of a handheld camera in certain scenes. My favourite example of such artificial artefacts is a blink-and-you-miss-it moment in *The Lego Movie* (dirs. Phil Lord and Chris Miller, 2014). Batman is whisking our heroes away from the baddies at high speed in the Batmobile, which sprays mud or water as it passes the (notional) camera. Being in the Lego universe, however, the debris that smudges the camera lens is formed of fuzzy, but recognisable, Lego pieces.

While in most of these examples there is a knowing wink to the audience, the blurring of boundaries between digital imaging and the conventions of analogue image capture can become increasingly bewildering. The critically acclaimed movie *Bait* (dir. Mark Jenkin, 2019) is filmed in black and white, and features extensive scratches and imperfections on the footage throughout the movie. While watching the film I had assumed these imperfections had been added on digitally as an atmospheric device, but during a Q&A session with the director at the end of the film this was revealed not to be the case. Jenkin shot the movie using a 1976 16 mm clockwork Bolex camera, and hand processed some 13,000 ft of film using an antique Bakelite rewind tank.[22] So while Jenkin's artistic decision to produce the film in this fashion inevitably meant there would be artefactual imperfections on the print due to the nature of hand processing the film, the imperfections can be considered "genuine" rather than artificial.

IN AND OUT OF FOCUS

In the case of both lens flare and the characteristic appearance of hand-processed film, a seeming imperfection or flaw of the image capture technology is utilised for stylistic or artistic effect. Likewise, film directors have also repurposed loss of focus in an image for similar ends.

Focus pull has provided directors with a means of signposting to audiences where their attention should lie, such as alternating between characters located in near-field and far-field positions.

However, the use of focus pull when only one character is within the shot can convey information to the audience about that character's state of mind or physical condition. For example, in *The Graduate* (1967) director Mike Nichols uses a dramatic and lengthy focus pull on Elaine (played by Katharine Ross) at the moment she realises that Benjamin (played by Dustin Hoffman) has been having an affair with her mother, Mrs Robinson. As the penny drops, her anguished face finally comes into sharp focus. Conversely, in the Russian World War II drama *The Ascent* (1977), director Larisa Shepitko used focus pull in the other direction (sharp focus to blur) to convey a soldier's fluctuating conscious level.

It is interesting to note the frequent use of focus pull in the majority of CGI movies as a specific, well-understood narrative device. If desired, every part of the screen in every frame could be in focus, but moviemakers deploy focus and lack of focus in a similar fashion to black/white, red-figure/black-figure, "on/off". An artefact of the image becomes another storytelling tool.

Multi-Oscar winning director Ang Lee discovered there can also be drawbacks to filming in crystal-clear focus in his 2016 film *Billy Lynn's Long Halftime Walk*. This was a landmark production, the first movie to be filmed at 120 frames a second (most films are 24 fps) and in 4K high definition digital 3D. In an interview, Lee described how he spent most of his time directing his lead actor Joe Alwyn, but could do little about extras positioned 50 yards away, normally smudged out of the scene, but now brought into sharp focus, overacting, or otherwise.[23]

The ultra-high definition offered by 21st-century technology may feel like a recent conundrum for film-makers, but active choices about what should and should not be in focus date back to the 1940s when director Orson Welles and cinematographer Greg Toland pioneered the use of "deep focus" in *Citizen Kane* (1941). This technique, achieved via a combination of faster film stock, more powerful studio lighting and some innovative use of lenses, allowed characters in both the near field and far field to both be in focus simultaneously. This innovation was heralded as liberating audiences, enabling the viewer the choice of where to direct their attention, but a subsequent backlash characterised the technique as gimmicky and self-conscious, critics pointing out that the director remains firmly in control of where the audience's attention is steered through the composition of each shot and the editing process.[24]

While deep focus has largely fallen out of favour, one of its methods – split dioptres – is still employed periodically. This consists of using a half circular convex glass attachment to the front of the camera, making half the lens near sighted while the other half remains far sighted. This allows a character to be positioned very close to the camera in one half of the frame and another character to be further away in the other half of the frame. While both characters are sharply in focus, anything in the middle ground between them is out of focus. Split focus shots have been employed by Steven Spielberg in *Jaws* (1975), Quentin Tarantino in *Pulp Fiction* (1994), and more recently by Luca Guadagnino in *Suspiria* (2018). The technique is a particular favourite of Brian De Palma, who has used it in nearly all his movies, including *Carrie* (1976), *The Untouchables* (1987), and *Mission: Impossible* (1996). In *Blow Out* (1981) De Palma uses split focus in some 15 separate shots.[25] A computer-generated version of this technique even appears in *Toy Story 4* (dir. Josh Cooley, 2019), alongside a host of other lens-mimicking tricks innovated by director of photography Patrick Lin.[26, 27]

In each case a split dioptre shot is used to heighten tension and provoke a feeling of unease in the viewer, as it is an unfamiliar appearance not replicated in our normal visual experience. It is interesting to compare this feeling of unease at this "unnatural" appearance – a clash of what we see on screen versus how we think things should appear – with the reaction of radiologists to CT images generated by the (relatively) new technique of iterative reconstruction. This is a data-processing technique that has allowed substantial radiation dose savings to be achieved through noise reduction in CT image processing. When this method was first introduced many radiologists complained that the images just didn't look right, describing the appearances as "plastic-like", "waxy", "pixillated", "paint-brushed", "blotchy", or "over-smoothed".[28, 29] One study evaluating image quality in relation to iterative reconstructions even utilised a 1–5 scoring system to specifically assess the "pixilated blotchy appearance" of CT images.[30]

In modifying the appearance of CT images to match the radiologists' perception of how they should look, it is difficult to know if the pictures get closer to or further from anatomical reality.

ARTEFACT VS ARTIFACT

Most dictionaries suggest the words "artifact" and "artefact" can be used interchangeably, the main difference being the former is conventionally used in North America and the latter is preferred in Britain. Such dictionaries do describe two different meanings:

1. An object made by a human being, typically one of cultural or historical interest
2. Something observed in a scientific investigation or experiment that is not naturally present but occurs as a result of the preparative or investigative procedure[31]

Both meanings can apply to either spelling. However, some internet discussion on this topic suggests that the archaeological meaning is increasingly being described by "artifact" and the interference type meaning is described by "artefact".[32]

Certainly, within radiology it can be useful to have different spellings for these separate meanings, as the two quite often converge. Take a look at Figure 3.16 – this is a transverse (or axial) MR image of a child's abdomen, but it is quite difficult to recognise it as such due to the big "black hole" (or "shark bite" as one of my surgical colleagues described it) obscuring much of the abdomen. This appearance is an example of susceptibility artefact, mentioned earlier in the chapter, and most commonly caused by ferromagnetic metal. It was therefore suspected the child might have previously ingested a metal object, and an abdominal radiograph (shown in Figure 3.17) subsequently proved this to be the case, a metallic density visible in the

FIGURE 3.16   A transverse (or axial) MR image of a child's abdomen, with a large area of susceptibility artefact obscuring much of the abdomen. Image provided by the author.

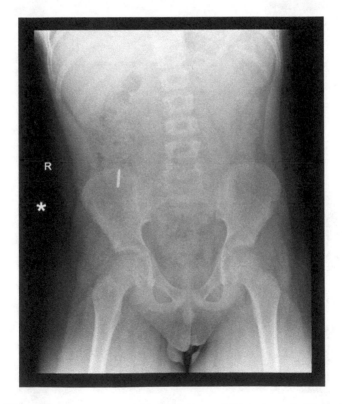

FIGURE 3.17   An abdominal radiograph (of the same child as shown in Figure 3.16) with a metallic density visible in the right lower abdomen, most likely a swallowed coin seen side-on. Image provided by the author.

right lower abdomen, most likely a swallowed coin seen side-on. In this example we have an artifact (the coin) causing imaging artefact (the black hole or shark bite).

We have already encountered some examples of motion artefact in young children. To prevent this when radiographs are acquired, parents or carers are sometimes instructed to hold a child still in the required position. Figure 3.18 shows a young child's hand being held in position by an adult carer, and Figure 3.19 is a skull radiograph in which the child is similarly being held steady. The adult fingers visible in the lower portions of the images can be labelled artefactual, but as with the anti-scatter grid we considered earlier they serve to improve the overall quality of the image. Putting technical considerations to one side the positioning of hand and face seen in Figure 3.19 triggers associations of affection and tenderness. Although the child may be being held for practical reasons the pose replicates that of a pre-kiss embrace, and perhaps we should not forget that parental love manifests itself as countless acts of practical assistance.

Parental love is writ large in Figure 3.20, a photograph of David Octavius Hill with his daughter Charlotte, taken in 1843. As a pioneering early photographer, Hill was well aware his tight embrace would serve to steady her for the long exposure time of several minutes, but to the modern viewer the embrace gains poignancy in

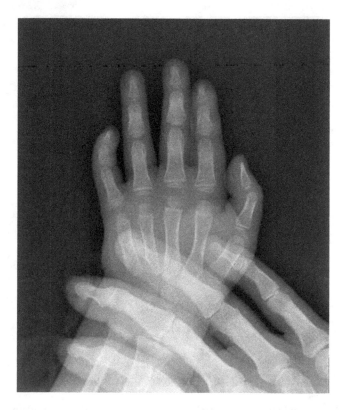

FIGURE 3.18   Plain radiograph, showing a young child's hand being held in position by an adult carer's hand. Image provided by the author.

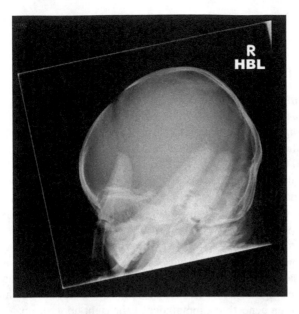

FIGURE 3.19 Skull radiograph in which the child's head is being held steady by an adult carer's hands. Image provided by the author.

FIGURE 3.20 Charlotte Hill and David Octavius Hill with his daughter Charlotte, photograph, 1843, by David Octavius Hill and Robert Adamson. Image credit: National Galleries of Scotland. Edinburgh Photographic Society Collection, gifted 1987.

knowing the child's mother had died two years previously and that Charlotte herself would die in her early twenties.[33,34]

LENDING A HAND

Fans of *The Addams Family* (variously a 1930s cartoon, 1960s TV show, 1990s feature films and a 2019 CGI movie) will recall the character Thing T. Thing (often just known as "Thing"), a dismembered forearm/hand or sometimes just hand with a life of its own. In the more recent animated movie *I Lost My Body* (dir. Jeremy Clapin, 2019) a severed hand journeys across Paris searching for the remainder of its body. We also have already come across a Thing-like dismembered hand in Figure 1.7, Chapter 1. Perhaps with all these disembodied hands being such a familiar visual motif, you will not be overly surprised by what you see in Figure 3.21. But I was certainly surprised. It is an illustration from Valverde de Hamusco's *Anatome corporis humani* from 1589, by the Spanish artist Gaspar Becerra. It shows an *écorché* figure (skin removed to reveal the muscular structures beneath) in a characteristic pose of Renaissance-era anatomy texts. What struck me as peculiar about this image is that the figure's right forearm has been removed and relocated to an entirely non-anatomical location beneath the remainder of the right upper limb.

Alongside this unusual image take a look at Figure 3.22. This is an image from an MRI examination of a baby, in which an effect known as wraparound artefact has resulted in the child's left arm (conventionally displayed on the right-hand side of the image as we look at it) being transposed to the other side of the image giving the impression that the child is holding hands with a twin. The two images are almost direct visual analogues, albeit with the wandering arm being located in different positions.

In the case of the *écorché* with the amputated hand, Becerra has made a deliberate choice to disassemble the figure, removing the hand to enable the entirety of the body to be displayed on a single page, albeit with continuity of the human frame interrupted. It is by no means an anomaly – within the text there are several other similar illustrations (a little like the variability in the Thing in different Addams Family adaptations, variable portions of the upper limb are removed in different illustrations). The deliberate nature of this appearance is underlined by an overlap in the coverage of the respective portions of the right upper limb, with the elbow region portrayed in both the floating forearm and the upper arm attached to the body. To the modern eye, this choice appears to be paradoxical. While the limitations to artistic vision imposed by the rectangular dimensions of the page on which the woodcut would be printed are understandable, it seems perverse to remove the hand so that the whole figure will fit onto a single page. Why not portray the figure in a different pose, with the arm not elevated so high? Why not shrink to fit the image so the whole thing fits onto the page instead of maiming the figure?

The answers are elusive, but I would speculate that the strong visual traditions of how anatomical figures should be portrayed weighed heavily upon the artist. In this case, portraying the figure in a particular pose has clearly superseded the consideration of portraying the figure as one continuous structure. Assuming the "shrink-to-fit" solution was unacceptable on the grounds that this would prevent details within

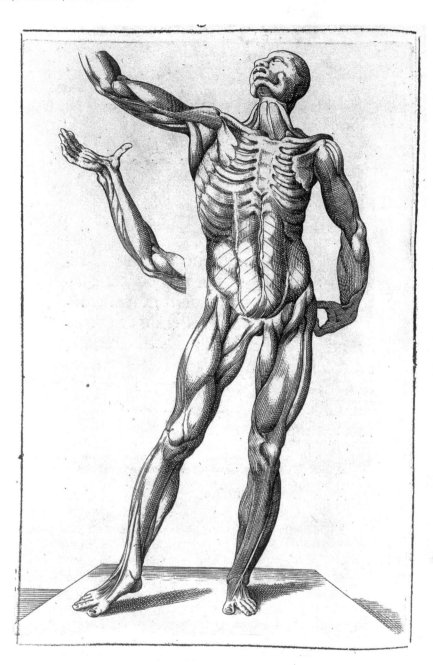

FIGURE 3.21 An illustration from Valverde de Hamusco's Anatome corporis humani from 1589, by the Spanish artist Gaspar Becerra. It shows an *écorché* figure in a characteristic pose of Renaissance-era anatomy texts. The figure's right forearm has been removed and relocated to an entirely non-anatomical location beneath the remainder of the right upper limb. Credit: Anatome corporis humani/avctore Joanne Valverdo. Wellcome Collection. Attribution 4.0 International (CC BY 4.0).

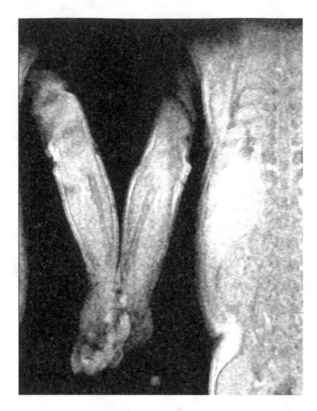

FIGURE 3.22   An image from an MRI examination of a baby, in which an effect known as wraparound artefact has resulted in the child's left arm (conventionally displayed on the right-hand side of the image as we look at it) being transposed to the other side of the image giving the impression that the child is holding hands with a twin. Image provided by the author.

the illustration being clearly visible, it would seem the convention of the heavenward gesture trumps the convention of maintaining the body as an intact structure.

In the case of wraparound artefact, the situation is admittedly rather different. The radiographer selecting the parameters which produce the image in an MRI scanner does not make an active decision to produce a wraparound appearance. However, a similar pragmatic limitation is imposed by the rectangular field-of-view, directly analogous to the rectangular page of the anatomical textbook, and an active choice does have to be made in terms of attempting to capture the most valuable anatomical and pathological information within that rectangle.

Wraparound (or aliasing) artefact occurs when the selected field-of-view is too small in relation to the selected body part, such that body tissue just outside the selected rectangle of interest produces radiofrequency signals which the scanner computer interprets as having originated from within the field of view. The aberrant data is shifted to the other side of the image such that patients' noses end up superimposed over the back of their heads, or babies appear to be holding hands with phantom twins. The problem can be avoided by using a larger field-of-view, but like the "shrink-to-fit" solution eschewed by Becerra, this would potentially reduce the spatial resolution of the resulting image.

In producing diagnostic quality images on an MRI scanner, radiographers are battling against the arbitrary dimensions of rectangular fields-of-view, the somewhat arbitrary conventions of anatomical planes, extensive variability in the size and proportions of the human frame as well as numerous unpredictable human factors. In a children's hospital such as where I work both patients and their families can be understandably anxious and apprehensive. The empathy and skill used by radiographic staff to reassure them, both prior to and throughout the scan, is truly impressive. Alongside the crucial people skills and the technical mastery of highly sophisticated imaging equipment, radiographers also need to exercise pragmatic judgement in relation to whether artefact is acceptable or not within a specific clinical examination. I have pointed out that radiographers do not intentionally look to produce wraparound artefact (or any of the other many artefacts which can potentially degrade MR images for that matter); however, they may need to decide whether a sequence in which a patient's nose has ended up in the back of their head needs to be repeated, or whether the images remain of diagnostic quality. It is possible, for example, that if the clinical question relates to the pituitary gland, located centrally, that wraparound artefact that has relocated the patient's nose into the back of their brain may not actually matter, provided it is not interfering with visualisation of the pituitary.

So it may be that, just as the artist made the active choice to cut off the *écorché's* hand and depict it below the forearm in order to maintain the conventional anatomical pose without compromising illustrative detail, the radiographer may decide to persevere with an image in which an individual's nose pokes into the back of their head in order to maintain the conventional anatomical plane without compromising spatial resolution. In both cases the constraints of the rectangular image format force choices to be made about what is an acceptable image within a given context.

LINEAR PROGRESSION

We will start this section with an excerpt from *Of Human Bondage* by W. Somerset Maugham:

> "What about the black line?" cried the American, triumphantly pushing back a wisp of hair which nearly fell in his soup. "You don't see a black line round objects in nature".
> "Oh, God, send down fire from heaven to consume the blasphemer", said Lawson.
> "What has nature got to do with it? No one knows what's in nature and what isn't! The world sees nature through the eyes of the artist. Why, for centuries it saw horses jumping a fence with all their legs extended, and by Heaven, sir they were extended. It saw shadows black until Monet discovered they were coloured, and by Heaven, sir they were black. If we choose to surround objects with a black line, the world will see the black line, and there will be a black line; and if we paint grass red and cows blue, it'll see them red and blue, and by Heaven, they will be red and blue".[35]

The premise of the quote above is that the outline (in this case around the nude female figure of *Olympia*, by Manet) is an entirely artificial device employed by the artist, but which if pulled off with sufficient skill can be perceived by the viewer as being entirely natural.

Instead of using an illustration of *Olympia*, which you can most likely conjure in your own head without even requiring a Google image search, I have chosen an example from another 19th-century artist who made emphatic use of the outline a deliberate stylistic device, Honoré Daumier.[36] *Les Spectateurs – Foyer du théâtre* (Figure 3.23) shows an exclusively male audience watching a play, with each figure, including hair, hats, and some impressively bushy moustaches all rendered with black outlines. To the modern viewer, the rather inscrutable facial expressions may

FIGURE 3.23   Les Spectateurs – Foyer du théâtre by Honoré Daumier (1888). An all-male theatre audience shown both sitting and standing as they gaze towards the stage. Image credit: Everett Collection/ Shutterstock

bring to mind the phrase "male gaze", although I do not know what they are watching on stage. Daumier's principal source of income was as a satirical cartoonist (his caricature of King Louis-Philippe, *Gargantua* led to a six months prison sentence in 1832)[37] and the cartoon-like appearance of figures in this and other of his paintings may not seem like a revolutionary advance in artistic technique. However, this unapologetic use of the outline influenced Manet, Matisse, and Gaugin, with knock-on inspiration for 20th-century innovators such as Picasso.[36] Daumier's oil paintings have also been lauded as the first impressionist works, and a source of inspiration to Van Gogh (take a look back at those shoes in Figure 3.1 – is that an outline we can see around them?).[38]

While drawing an outline of an object or figure is probably the most common way to get started on a representational planar image, most of us would recognise the artificial nature of this method. We know there is not a continuous (most commonly) black line delineating the margins of any particular object in space. Indeed, it is in confounding our expectations that makes Daumier and Manet's use of an outline intriguing. However, the outline is perhaps not entirely artificial. As we saw in Chapter 2, the human visual system employs a form of edge enhancement which if not quite producing the type of thick black line around every object that we see delineating *Olympia* or those bushy moustaches, does mean edges between and within objects do appear to have more discrete margins than they do "in reality".

"So what?" you might ask – a quirk of the human visual system does not mean that the use of an outline gets us closer to the underlying truth of reality, even if the technique is perhaps more closely aligned to how we see the world than we first supposed. If we were to imagine an individual whose visual system had a more pronounced edge enhancement than the rest of us, it is not such a leap of physiological or genetic imagination that such a person really might view the world with black lines around objects and figures, seeing the world in a similar fashion to a cartoon or comic book. Yet the reality purists would still point out that this is an idiosyncrasy of human vision, not the truth of underlying reality. In which case let us consider MRI sequences known as gradient echo or opposed-phase images in which a characteristic appearance known as "India ink" artefact is to be found.

In these sequences any voxel in which sufficient quantities of both fat and water are to be found leads to the signal from each of these substances cancelling one another out, and the corresponding pixel will be displayed as black. Within the human body, interfaces between fatty tissue and tissue or structures with a high water content are common, and such interfaces commonly delineate specific structures such as organs and so forth.[39] This sequence, therefore, produces black lines around any organ or structure surrounded by fatty tissue as you can see in Figure 3.24. The distinctive black delineation of the kidneys is hard to describe as anything other than an outline. Figure 3.25 is another example of this appearance – this is the same child shown in Figures 3.16 and 3.17 after the coin had been passed, with the "shark bite" no longer seen.

So here we have a scientifically robust technique which has generated an outline of objects in an accurate, objective fashion. Now, one can certainly argue that this is only a single MRI sequence out of a range of 30 or so well established in clinical practice, and you can point out that this India-ink appearance only outlines some selective interfaces. You could also point out that the black line we are seeing is artefactual,

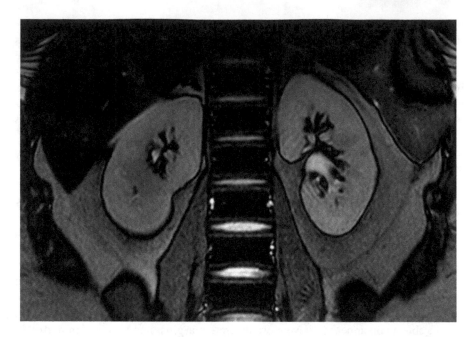

FIGURE 3.24 MRI scan of the abdomen in coronal plane, demonstrating "India ink" appearance. Notice how the kidneys are outlined by distinctive black lines. Image provided by the author.

because of the size of the voxel. If the spatial resolution was sufficiently high (so that the voxels were sufficiently small) we would be able to overcome this effect because all the fat would be confined to voxels on one side and all the water would be confined to voxels on the other side. Perhaps a few pesky voxels might continue to have a bit of both, resulting in a scattering of black pixels in the region of the interfaces, but if spatial resolution is sufficiently high, the India-ink lines will by-and-large disappear.

Critics of the outline – whether drawn with a pen or pencil by an artist, or produced by MRIs as stipulated by objective physical criteria – can legitimately point out that the outline is not real. However, if we are willing to accept that the MRI is a legitimate *representation* of reality, then it becomes increasingly difficult to argue that the artist's use of an outline is not also a legitimate representation.

There is, it turns out, a twist in the tale. I have made use of the seemingly objective, science-driven MRI appearances to justify and add respectability to the artist's use of black outlines. Figure 3.26 shows one of these MRI black lines at the highest level of magnification the computer will allow me to. At this level of magnification, I would expect to be able to see individual pixels, with a uniform tone of grey within each pixel, each of which should occupy a square on the screen (the yellow outlined square located centrally within the image indicates the size of an individual pixel). Instead, there are no sharp divisions of tone and no obvious squares. So what's going on? The imaging software that displays the MRI examination on the computer uses an algorithm to smooth off the sharp edges of individual pixels, resulting in a more "natural" looking appearance.

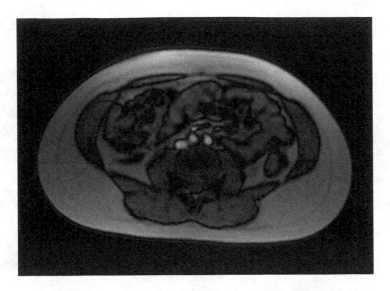

FIGURE 3.25   MRI scan of the abdomen in axial plane. This is the same child shown in Figures 3.16 and 3.17 after the coin had been passed, with the "shark bite" no longer seen. Image provided by the author.

FIGURE 3.26   A segment of Figure 3.25 at a high level of magnification. The yellow outlined square located centrally within the image indicates the size of individual pixels. Image provided by the author.

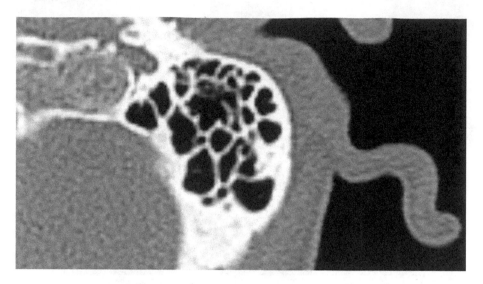

FIGURE 3.27   Magnified segment of a CT head scan showing the mastoid air cells. Image provided by the author.

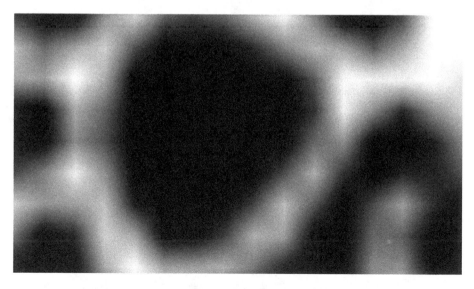

FIGURE 3.28   The central portion of the CT scan shown in Figure 3.27 at the highest level of magnification. We are looking at low-density air cells (in black) surrounded by very thin walls of bone, shown in white. Image provided by the author.

Figures 3.27 and 3.28 show a similar process at work in relation to CT images. Figure 3.27 shows a zoomed-in section of a head scan showing the mastoid air cells (part of the skull just next to the ear), and Figure 3.28 shows the central portion of this image at the highest level of magnification. We are looking at low density air cells (in black) surrounded by very thin walls of bone, shown in white. Although we can discriminate the location of individual voxels more readily than on the MRI example above, more than one grey shade is displayed within individual voxels demonstrating another smoothing algorithm at work. The display of the "raw data" has been manipulated with an expectation of what the viewer is anticipating seeing. So, these apparently scientific images utilise just the same type of visual sleight of hand that would give Photoshop manipulation or Instagram-style photo filters a bad name.

SUMMARY

The business of untangling what should and what should not be within an image to produce a "true likeness" of the subject is a little more complicated than we might first suppose. In this chapter, we have seen how artists in a variety of traditions may seek to either accentuate or eliminate the mark of the artist in their work. In art, movies, and radiology alike, artefacts which some would argue should not be within the image have proven to have either artistic or diagnostic merit, resulting in not only the purposeful inclusion of existing artefact, but also the deliberate fabrication of such artefacts. While in medical imaging the predominant emphasis is on eliminating artefact as a potential source of confusion or misdiagnosis, we have also considered instances in which what could be classified as artefact actually provides clinically useful information.

In both the artistic and medical arenas, the distinction between what is a "true" likeness of the subject of interest (or in the case of abstract art a true representation of the intended artistic vision) and what is a corruption or distortion becomes rather blurry. The use of manufactured imaging artefact in movies, videogames, and even within medical imaging in the pursuit of (apparent) verisimilitude highlights this ambiguous relationship. Distinctions between artifact, artefact, and artifice become increasingly hazy.

In *Art and Illusion*, E.H. Gombrich describes some visitors looking around Henri Matisse's studio. A lady is said to have commented, "the arm of this woman is surely far too long!", to which Matisse replies, "You are mistaken, Madame. This is not a woman, this is a picture".[40]

In short, pictures (whether painted, photographic, or radiographic) usually have more in common with other pictures than they do with their intended subject matter.

REFERENCES

1. *Video Transcript*: Bridget Riley in conversation with Sir John Leighton www.nation algalleries.org/sites/default/files/transcriptions/Video%20Transcript%20Bridget%20Riley.pdf, last accessed 17/07/2021.

2. F. Spalding, Bridget Riley and the poetics of instability, in *About Bridget Riley: Selected Writings 1999–2016*, D. Globus and K. Schubert (eds), Ridinghouse, London, 2017, p 21.
3. J.P. Hodin, Vincent Van Gogh as quoted: The painter's handwriting in modern French art, *The Journal of Aesthetics and Art Criticism*, 7(3), March 1949, 181–199.
4. Loving Vincent, http://lovingvincent.com/the-paintings,2,pl.html, last accessed 17/07/2021
5. Van Gogh Museum, www.vangoghmuseum.nl/en/stories/inspiration-from-japan#0, last accessed 17/07/2021.
6. Wikipedia, Japonaiserie (Van Gogh), https://en.wikipedia.org/wiki/Japonaiserie_(Van _Gogh), last accessed 17/07/2021.
7. D. Hockney and M. Gayford, *A History of Pictures*, Thames & Hudson, London, 2016, p 39.
8. The Clark Museum, www.clarkart.edu/Museum/Publications/Painting-and-Sculpture/ Like-Breath-On-Glass-Whistler,-Inness,-And-The-Ar Last accessed 17/07/2021.
9. D. Hockney and M. Gayford, *A History of Pictures*, Thames & Hudson, 2016, p 51.
10. Tate, Readymande, www.tate.org.uk/art/art-terms/r/readymade, last accessed 17/07/ 2021.
11. New York Times, Is it art? Eyeglasses on museum floor began as teenagers' prank, www. nytimes.com/2016/05/31/arts/sfmoma-glasses-prank.html, last accessed 17/07/2021.
12. R.F. Farr and P. J. Allisy-Roberts, *Physics for Medical Imaging*, Saunders, London, 1998, p 206.
13. E.M. Haacke, S. Mittal, Z. Wu, J. Neelavalli and Y.-C.N. Cheng, Susceptibility-weighted imaging: Technical aspects and clinical applications, Part 1, *American Journal of Neuroradiology*, 30(1), January 2009, 19–30, doi:10.3174/ajnr.A1400
14. Cambridge in Colour, Understanding camera lens flare, www.cambridgeincolour.com/ tutorials/lens-flare.htm, last accessed 17/07/2021.
15. Conrad L. Hall, 1926–2003, *American Cinematographer*, https://theasc.com/magazine /may03/cover/page2.html, last accessed 17/07/2021.
16. Pilgrim Akimbo, Thinking of you Conrad Hall, https://pilgrimakimbo.wordpress.com/ 2007/02/07/thinking-of-you-conrad-hall-a-random-observation-about-a-film-he-didnt-shoot/, last accessed 17/07/2021.
17. Polygon, J.J. Abrams says his wife convinced him to stop with the lens flares by Samit Sarkar, www.polygon.com/2016/3/11/11206910/jj-abrams-lens-flare-stop-star-trek-into-darkness, last accessed 17/07/2021.
18. The Verge, J.J. Abrams apologizes for overusing lens flare: 'I know it's too much' by nateog, www.theverge.com/2013/9/30/4788758/j-j-abrams-apologizes-for-his-overu sing-lens-flares, last accessed 17/07/2021
19. S.M. Hugh, X-Ray visions: Radiography, "Chiaroscuro", and the fantasy of unsuspicion in "Film Noir", *Film Criticism*, 32(2, Winter), 2007–2008, 2–27 (p 4).
20. Ibid p 10.
21. Ibid p 5.
22. Bait, www.baitfilm.co.uk/how-we-made-it-1, last accessed 17/07/2021.
23. The Guardian, Interview: Ang Lee: 'I know I'm gonna get beat up. But I have to keep trying' Ryan Gilbey, *The Guardian*, 3/10/2019, www.theguardian.com/film/2019/oct/ 03/ang-lee-i-know-im-gonna-get-beat-up-but-i-have-to-keep-trying, last accessed 17/07/2021.
24. Roland Denning, The return of deep focus? *Red Shark News*, www.redsharknews.com/ production/item/3010-the-return-of-deep-focus-shallow-depth-of-field-is-not-the-only-way, last accessed 17/07/2021.

25. Mike Bedard, The split diopter lens explained with eye-popping examples. *Studio Binder*, 17/05/2020, www.studiobinder.com/blog/split-diopter-lens/, last accessed 17/07/2021.

26. The Nerdwriter, The real fake cameras of toy story 4, https://youtu.be/AcZ2OY5-TeM, last accessed 17/07/2021.

27. V. Renée, How Pixar's 'real' fake cameras make their movies look so realistic. *No Film School*, https://nofilmschool.com/toy-story-4-pixar-cinematography, last accessed 17/07/2021.

28. L.L. Geyer, U.J. Schoepf, F.G. Meinel, J.W. Nance, Jr, G. Bastarrika, J.A. Leipsic, N.S. Paul, M. Rengo, A. Laghi, C.N. De Cecco, State of the art: Iterative CT reconstruction techniques, *Radiology*, 276(2), 2015, 339–357. doi: 10.1148/radiol.2015132766. PMID: 26203706.

29. Atul Padole, Ranish Deedar Ali Khawaja, Mannudeep K. Kalra, and Sarabjeet Singh, CT radiation dose and iterative reconstruction techniques, *American Journal of Roentgenology*, 204(4), 2015, W384–W392.

30. B. Pauchard, et al. Iterative reconstructions in reduced-dose CT: Which type ensures diagnostic image quality in young Oncology patients? *Acad. Radiol.* 24(9), 2017, 1114–1124, ISSN 1076-6332, doi:10.1016/j.acra.2017.02.012.

31. Google, Definition from Oxford languages, www.google.com/search?q=artefact+de finition&rlz=1CACCCC_enGB868&oq=artefact+de&aqs=chrome.0.0j69i57j0l3j6 9i60l3.7169j1j7&sourceid=chrome&ie=UTF-8], last accessed 17/07/2021.

32. English Language and Usage, Discussion forum, https://english.stackexchange.com/q uestions/37903/difference-between-artifact-and-artefact, last accessed 17/07/2021.

33. C. Henderson, *The Book of Barely Imagined Beings: A 21st Century Bestiary*, Granta Books, London, 2012, p 220.

34. S. Stevenson, *Facing the Light: The Photography of Hill & Adamson*, National Galleries of Scotland, Edinburgh, 2002, p 11.

35. W. Somerset Maugham, *Of Human Bondage* Loc 2892.

36. J.P. Hodin, The painter's handwriting, in *Sign, Image, Symbol*, G. Kepes (ed), G. Braziller, New York, 1966, p 152.

37. National Gallery, www.nationalgallery.org.uk/artists/honore-victorin-daumier, last accessed 17/07/2021.

38. Honoré Daumier by Jean Adhémar, www.britannica.com/biography/Honore-Daumier, last accessed 17/07/2021.

39. V. Indiran and V. Sivakumar, Diagnostically useful MRI artifact – India ink artifact, *Neurol India*. 67, 2019, 573–4.

40. E.H. Gombrich, *Art and Illusion: A Study in the Psychology of Pictorial Representation*. 5th edition, Phaidon Press, London, 1977, p 98.

4 Maps, Mirrors, and Manipulation

Figure 4.1 is one of the first radiographs made by Wilhelm Röntgen in late 1895. It doesn't require any radiographic knowledge to recognise that it shows a compass, but it does help introduce us to an important radiographic concept. The needle of the compass is aligned roughly along the top-to-bottom or longitudinal axis of the print, which corresponds to the north-south axis indicated by the compass markers. It would not be an unreasonable assumption to suppose that the tip of the needle closest to the top of the image is pointing north and the needle tip pointing downwards in the image is directing us south. However, it would be exactly that – an assumption, as without knowing the magnetic polarity of the needle (most commonly indicated by red colouring for the north half on modern compasses) we cannot be sure that the markings on the circumference of the compass are correctly aligned.

In radiographic practice, a similar uncertainty is encountered in relation to which side of the body is being imaged. For the majority of radiographic examinations (and indeed most medical imaging examinations in general) it can be impossible to tell with certainty which side of the body has been imaged unless the radiographer has provided a left/right marker on the image (typically a prominent "L" or "R" located in the top corner of the radiograph). In the context of hard copy films on light boxes, the radiograph can easily be displayed the wrong way around if reference is not made to the side marker and on a computer monitor the left/right orientation of the image can be readily reversed at will. Making sense of which side is which is therefore of fundamental importance within clinical imaging, but to do so we shall venture a little further afield than the X-ray department.

ORIENTATION

In *A History of the World in Twelve Maps*, Jerry Brotton demonstrates how maps – and world maps in particular – often reveal as much about the political and cultural environment in which they were created as they do about the terrain they are intended to portray. This is well demonstrated by a strong tendency throughout history for cultures to situate their own country or most valued religious site at (or at least within suspiciously close proximity to) the centre of the map. A degree of cartographic uniformity does exist insofar as the large majority of large-scale maps originating from all cultures have employed axes based on the four cardinal compass directions – the east-west axis long recognised by the rising and setting of the sun, the north-south axis inferred by the position of the North Star or the midday sun (prior to the introduction of the compass). However, the specific alignment of the map (i.e. which compass direction is located at the top of the map) has varied between cultures.[1]

DOI: 10.1201/9780367855567-4

FIGURE 4.1   A compass whose magnetic needle is entirely surrounded by metal, viewed through X-ray. Photoprint from radiograph by W.K. Röntgen, 1895. Credit: Wellcome Collection. Attribution 4.0 International (CC BY 4.0).

Very few examples of maps with west at the top are known to exist because of the negative connotations of the sun disappearing in this direction (also said to be the direction in which Christ was facing while being crucified).[2] Until the 15th century, the majority of Judaeo-Christian maps put east at the top and west at the bottom, following in the tradition of polytheistic sun-worshipping cultures that revered east as the direction of life and renewal. Indeed, the origin of the word orientation comes from *oriens*, the Latin word for east. In ancient Hebrew, the words for north (*smol*) and south (*teman*) also mean left and right respectively, implying an eastward-facing worldview.[3] Churches have traditionally been aligned on an east-west axis with the altar located at the east end of the building.[2] The direction of the congregation facing towards the main altar has become known as "liturgical east", regardless of what the actual orientation might be. So for example, the guidebook to the Anglican cathedral in Liverpool refers to "west doors" and the "the great east window" when these actually face north and south respectively.[4] Construction on the cathedral, the largest in the UK, began in 1904, by which time the imperative for churches being aligned in an east-west axis was less keenly felt. It is interesting to note that while on a spiritual level an eastward-facing direction is most valued, on the more prosaic consideration of which end of town to live the axis is reversed in the British Isles – the west traditionally being the desirable end of town in most UK cities due to prevailing wind conditions, with pollution and unpleasant odours from factories being blown towards the east.[5]

Although east was also revered to some extent within Islam, the direction of Mecca (*qibla*, "sacred direction") held even greater importance to Muslim cartographers. The rapid expansion of Islam into territory located north of Mecca in the 7th and 8th centuries meant that for those communities the *qibla* was due south, and consequently most Muslim world maps have historically been oriented with south at the top.[6] Figure 4.2 shows an example; north is located at the bottom, with west to the right surrounded by unknown seas. The Indian Ocean is shown on the left, with China, India, and Iran in boxes to the right.

In traditional Chinese maps the orientation is reversed, and – in common with most contemporary maps – north is located at the top of the map. This convention was based on the concept of the Emperor surveying his subjects in the south from a northerly location. As the Emperor is located in the north, the north is assigned as the top of the map, but geographically the emphasis is on the southern vista – the direction of warm winds and productive farmlands. Chinese houses have tended to face south for similar reasons. While historic Chinese maps look like they are "the right way up" to contemporary Western eyes, it is worth noting that the Chinese compass was oriented to point southwards (known as a *zhinan* – "southern pointer").[7]

On the one hand, then, making sense of geographic axes involves colossal physical entities such as the molten iron core within the Earth's crust (determining the north-south axis or at least the direction a compass needle points), and the motion of the Earth relative to the sun (determining the east-west axis). On the other hand, each of these cultures/religions has included a human dimension of sorts in determining the orientation of the map axes – the direction Christ faced on the cross, the birthplace

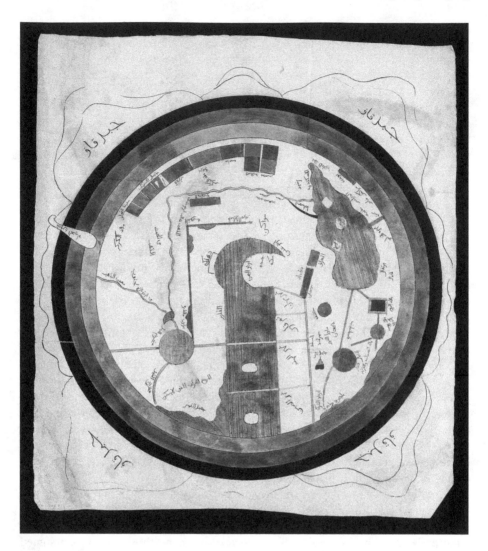

FIGURE 4.2 Islamic map of the world. The north is at the bottom, with the west to the right surrounded by unknown seas. The Indian Ocean, with the Red Sea, is on the left, with China, India, and Iran in boxes to the right. The other sea shown is the Mediterranean next to which is a black square indicating Rome and a circle Constantinople. The Nile flows from the Mediterranean to the east and into the north, then turns east and heads towards a large circle indicating its source in Africa. Credit: Wellcome Collection. Attribution 4.0 International (CC BY 4.0).

of Muhammad, and the direction by which the Emperor regards his subjects. The human dimension is particularly resonant in relation to the Chinese Emperor. Facing south, the Emperor has his back to the north. The Mandarin words for both north (Běi) and back (Bèibù)[8] share a common origin, and the fusion of anatomy and geography is also revealed by the Mandarin word for "recite" (Bèisòng), which originated from a pupil turning their back to their teacher (and presumably the blackboard or equivalent with any useful clues) to perform a text from memory.[7] The east-west axis, determined by the spinning of the Earth on its axis – vast and unconcerned by action occurring at a human level, takes on a very human scale in the Chinese tradition; from the perspective of the Emperor left is east and right is west.

This Emperor-centric framework for aligning left and right chimes with what linguists often refer to as the "egocentric" or "relative" framing of left–right orientation.[9] In the English language (and indeed most languages) spatial relationships are typically described from the point of view of the describer or by the describer putting themselves in the shoes of the individual to whom they are describing. This works fine so long as there is a common understanding of what direction the observer should be facing, but (as we shall see) can become problematic when this is more ambiguous. The language of Guugu Yimithirr, or Guguyimidjir, spoken by a few hundred Aboriginal Australians in Northern Queensland, offers an alternative means of describing spatial relationships. There are no words for left or right – speakers instead use the equivalent of the cardinal compass directions to describe where things are located. Rather than having left and right arms, then, you might have north and south arms, but of course by turning 90 degrees they become east and west arms and by turning a further 90 degrees your north arm has become your south arm and vice versa.[10] For a people highly attuned to their environment, intuitively familiar with geographical directions, this arrangement works highly effectively. For the rest of us, particularly those of us living in places where the sun (let alone the night stars), may not be visible for days at a time, we are stuck with getting to grips with our left and our right.

THE L-R PROBLEM

Until around 550 million years ago the left-right conundrum (referred to as "the L-R problem" by developmental biologists) had no basis in biological reality. That changed with the emergence of organisms such as *Ikaria wariootia*, a primitive organism the shape of a jelly bean, discovered in South Australia, and reported in 2020.[11] Measuring between 2 and 7 mm, *I. wariootia* is possibly the earliest example of a bilateral organism, with fossil evidence showing features consistent with a front and a back, a plane of symmetry resulting in a left and a right side, and a gut that opens at each end.

From such humble origins emerged all the rich diversity of asymmetric body plans within the animal kingdom, including our own. While in human anatomy this asymmetry is not readily apparent from the surface, some striking examples of pronounced body asymmetry exist elsewhere in the natural world. The male fiddler crab has a major claw or pincer much larger in size than on the other side, sometimes wider than its body, which is used in a waving display as a courtship ritual (the female crabs have symmetrically sized pincers).[12] *Perissodus microlepis* is a species of cichlid fish that feed on the scales of larger species in Lake Tanganyika. Around half of these fish have jaws twisted to the left side, which makes it easier to consume scales from the right flank of the donor fish. The remainder of the population have jaws twisted to the right, and feed from the left flank.[13] In a similar adaptation to facilitate feeding, *Pareas iwasakii*, a species of snake found in the southern Ryukyu Islands of Japan, has asymmetric jaws which makes it easier to eat snails with dextral (clockwise coiled) shells. This anatomical adaptation evolved as the large majority of snails have clockwise or right-sided shells, and snails with a left-sided/sinistral/anticlockwise shell are less likely to be eaten by this snake as a result.[14] However, being a "lefty" snail poses other challenges.

In the UK almost all garden snails have right-sided shells, and the mechanics of snail reproduction mean that snails must have their shells coiling in the same direction to be compatible. On this basis, the discovery of a rare sinistral snail by a retired scientist in London prompted researchers at the University of Nottingham to launch a media-backed international search for another left-sided snail to act as a mate. The snail was nicknamed after Jeremy Corbyn, leader of the Labour party at the time, and well known for his leftwing political outlook – a reminder that left and right have been subject to political associations of a similar nature to the cultural and religious associations of compass directions. "Jeremy the lefty snail" caused something of a media sensation and consequently not one but two sinistral snails were found and sent to Nottingham as potential mates. In a soap-opera style love triangle twist, these two snails promptly mated successfully with one another, producing 170 dextral offspring, but one did eventually also mate with Jeremy, producing 56 offspring, also all with right-sided shells. Jeremy died in October 2017, but not before helping raise the profile of biological left-right asymmetry in the public consciousness.[15]

The University of Nottingham's interest in Jeremy was in regard to unravelling the genetics of left-right asymmetry, no easy business with over 100 genes having been reported to be involved in left-right patterning in model organisms.[16] At least three of these genes, *Nodal*, *Lefty-1*, and *Lefty-2*, have been shown to be crucial in determining the left-right axis in mouse models via a mechanism known as the "nodal flow model". These genes are expressed in cells located at the embryonic L-R organising site (known as the node), and have the effect of generating unidirectional flow in motile monocilia (which all rotate in a clockwise direction).[17] This flow, in the direction of what will become the left side of the animal, enables the transport of cell signalling mechanisms (a "left-specific morphogen")[18] producing downstream asymmetric cellular activity, and ultimately the asymmetry in the final body morphology.[16,17,18]

While largely hidden beneath the skin surface in human anatomy, there is profound left-right asymmetry in the placement of the heart and major vessels, the lobar structure of the lungs and in the location of both solid and visceral organs in the abdomen. Deviations from the usual pattern of heart, stomach, and spleen on the left, liver on the right are rare, which most likely explains the first report of a laterality defect emerging relatively recently in around 1600 from the Italian anatomist and surgeon Girolamo Fabrizio. In 1788 Scottish physician and pathologist Matthew Baillie described for the first time the mirror-image reversal of the thoracic and abdominal organs, now known as *situs inversus totalis*.[18] This arrangement has no harmful consequences – it is reported that the oldest documented European, a woman alleged to have lived to 126, had this condition,[17] although I have been unable to independently verify this. However, heterotaxy syndromes in which there is partial reversal of the usual body plan (such as dextrocardia, in which the heart is located in the right side of the chest rather than the left, but with abdominal organs maintained in the usual configuration) are associated with potentially harmful structural abnormalities such as congenital heart defects.

Dextrocardia is relatively rare (with an incidence of 1 in 12,000)[19] but it has a tendency to appear quite frequently in radiology examinations. Prior to the use of digital display of images for the Royal College of Radiologists final exam, the wilier examiners are said to have handed chest radiographs showing dextrocardia to the candidates to fix onto the film viewer themselves. The more astute candidates would notice the side markers and correctly position the radiograph showing reversal of normal cardiac anatomy, with the heart sited to the right of the chest instead of the left. Others may not have noticed this and struggled to spot any abnormality until their error was pointed out. Fans of the medical sitcom *Scrubs* will be familiar with this scenario – the title sequence features a chest radiograph on the title image, with the heart located to the right of the chest. Internet speculation ensued as to whether this was an unintentional error, a reference to the medical inexperience of the shows' main characters, or that the image is displayed correctly but that the patient has *situs inversus*.[20]

In the era of hard copy radiographs the potential to cause harm to patients in the clinical environment as a result of this type of image reversal was well recognised, with a consequent emphasis on meticulous labelling on the part of the radiographer performing the examination, and label checking on the part of the radiologist. Radiographers would use radio-opaque "L" and "R" markers to indicate the side of the body, placed at an appropriate site within the area of exposure. These markers would often be accompanied by the initials of the radiographer in smaller letters, providing a signature of their handiwork. Since the advent of digital radiography the requirement to use radio-opaque side markers has diminished as the radiographer can mark the side as a digital imprint at the time of processing the image. This has the potential advantage of being able to select the location of the marker after the image has been taken, and thus being sure that the marker does not obscure any clinically important structures. However, some radiographers have suggested the "shoot first, label later" approach risks potential side-marking errors and have advocated continuing to use radio-opaque markers. They also argue that using initialled markers may consolidate a sense of pride in the radiographers' technical prowess, knowing that anyone viewing the image can trace its authorship, calling this the "signed artwork effect".[21]

As a brief aside related to the transition from hard copy to digital radiography, the notation used to indicate a fracture on A&E trauma films is worth a mention. In the UK, the tradition in the hard copy era was that if the radiographer spotted a fractured bone (or other significant findings) at the time of taking the radiograph, they would stick a small red dot to the film to alert the emergency team. In the digital era, this is no longer possible and most X-ray reporting monitors are greyscale in any case. As an alternative, many radiographers now use an asterisk (*), but others use characters to type out "RED DOT" on the image (only one character less than writing "FRACTURE"). Perplexingly, the phrase "red dot" has been registered as a trademark by a company producing artificial intelligence software for one of its algorithms.[22] If the computers themselves don't put me out of a job (more of which in Chapter 8), perhaps constraints on the vocabulary available to construct my reports will do.

THE RADIOLOGICAL RIGHT

With respect to most radiographs the image is typically displayed in such a fashion as to mimic the position of the patient being sat or stood in front of the individual looking at the radiograph. That is to say, the right-hand side of the patient is displayed on the left-hand side of the image and vice versa. In our own department there is an exception to this convention – radiographs of patients with curvature of the spine in whom surgery is being planned to correct their scoliosis have their images displayed in the opposite way to the normal convention (right side of the patient on the right side of the image etc.). The use of side markers and the ability to flip the image over on a mouse click mean confusion over which side is which is rare, but not unheard of. An emphasis on meticulous labelling and awareness of that labelling is therefore still very much required in the digital era. While this change to convention is potentially confusing for radiologists the rationale is that displaying the images in this fashion, as though seen from behind the patient rather than in front, is helpful to the spinal surgeons as it matches their view when operating.

Outside of the radiology department, the instilled association of "left-on-the-right" has the potential to cause confusion. I remember a former colleague describing her husband's withering put down after a wrong turn was taken while navigating in the car – "Did you mean the *radiological* right dear?". When looking at cross-sectional imaging such as CT and MRI, the images also follow the convention of the right side of the patient on the left-hand side of the image, but this is not quite as intuitive as imagining the patient sat or stood in front of you. In the case of axial images, the situation is more akin to the patient lying flat on their back in bed with the viewer located at the foot of the bed looking towards the head at the far end. However, whilst the point of view of the viewer is looking "upwards" towards the head end, the images are arranged such that the more superior (head end) images are displayed first and the inferior (feet end) images are displayed last. The ability of computers to rapidly display images sequentially means most radiologists will scroll back and forth between different sections of the body to look at individual structures, such that the "direction of travel" becomes somewhat arbitrary. However, even though I am well aware that conceptually patient's feet should be on my side of the computer monitor and their head on the far side of the monitor, I find it difficult to shake the feeling that as I scroll through a CT dataset that I am looking "down" on slices of the body rather than looking "upwards".

The left-right problem arises again in the context of antenatal imaging. Babies in utero do not tend to position themselves in line with conventional anatomical planes, and particularly in the context of multiple pregnancies it can be a major challenge to work out what side of which body is in view. One of my colleagues has a toy meerkat with "L" and "R" labels attached to help make sense of which side is displayed when reporting antenatal MRI scans, moving the meerkat in synchrony with the plane of each imaging sequence in order to not lose track of which side is which. I did suggest that using the Guguyimidjir system of labelling parts of the body based on compass directions might dispense with the need for the meerkat, but this advice did not go down particularly well.

LEFT UPPER LIGHTING

As we saw earlier in the chapter, churches have traditionally been aligned east to west, mosques aligned with reference to the direction of Mecca and Chinese houses built to face to the south. European painters, by contrast, have traditionally been advised to have their studios facing north.[23] This seems to be linked to conventions of directional lighting in artistic composition. Art historians have long suspected that there is a strong preference for artists to light scenes from the left-hand side, most commonly from the left upper position as the viewer inspects the work. A typical example of such a lighting scheme is shown in Figure 4.3, *The Anatomy*

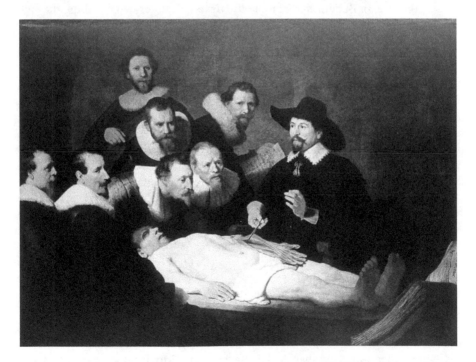

FIGURE 4.3 The anatomy lesson of Dr Nicolaes Tulp, by Rembrandt, painted in 1632. Several men are gathered around a cadaver as Dr Tulp demonstrates anatomical features of a partially dissected arm. The painting is lit in characteristic "upper left" style. Credit: Wellcome Collection. Attribution 4.0 International (CC BY 4.0).

Lesson of Dr. Nicolaes Tulp, by Rembrandt, painted in 1632. This suspicion was recently confirmed by a research paper in which 9,469 paintings were analysed, covering from 1500 BCE to 2000 CE, demonstrating that the left upper position for directional light was strongly favoured in the period 1420–1900.[24] Various reasons for this have been suggested including the idea that as most artists are right handed, a light source positioned from the right would cause a shadow of the hand on the canvas as they worked.[23, 25] Another theory is that in the same way written text is read from left to right for the majority of languages, paintings are also "read" from left to right in a similar fashion.[26] The direction of text would also appear to be related to most people being right handed, although it is worth noting that a number of major languages do not follow this convention, including Arabic, Hebrew and Urdu. Certainly, the use of left and right-sided symbolism is deeply embedded within Western art and culture as explained in *The Sinister Side* by James Hall.[27] Both Aristotle and Ptolemy aligned the right side of the body with the east, characterising it as courageous, strong, and masculine, and the left side of the body with the west (tender, effeminate, and secretive).[28] Plato viewed ambidextrousness to be the ideal state, with the prevalence of right-handedness being imposed by cultural factors, but in describing judgement in the afterlife the righteous souls pass to the right and the damned to the left, with a similar fate described in Christianity and in Islam. Likewise, in Buddhism, the route to Nirvana is described as taking the right path rather than the left.[29]

MIRRORS

A little while ago I noticed something was wrong as I looked at my reflection in the bathroom mirror. The embroidered writing on my T-shirt was disconcertingly easy to read. The text was not left-right inverted as would usually be expected. It took a moment for the penny to drop. My shirt was on inside out. I retraced my day to work out the implications. I'd only had the T-shirt on since I got home from work, having changed out of work clothes as part of coronavirus risk reduction measures. I think the only person outside my household who might have spotted my clothing faux-pas is the person serving me at our local Thai takeaway, but I had a coat on at the time. I think I got away with it.

Looking in the mirror is such a familiar, everyday experience that it takes a moment like this to remind us how artificial a mirror image actually is. In various artistic movements in which a conscious effort has been made to produce more naturalistic or life-like images, the phrase "holding up a mirror to nature" is often employed, as though a mirror image is the ultimate goal of the artistic endeavour. Yet the mirror image is a very specific, and in some respects a very peculiar representation of reality. While the image is left-right inverted, it is not top/bottom inverted. The front/back dimension (or z-axis) can also be conceptually challenging – I learnt this whilst attempting my own lockdown haircuts. Moving the scissors forward brought them closer to the mirror. On intuitive grounds this should move them closer to my reflection and the tuft of hair I was aiming for, but the scissors were in front of my own head and – of course – that of my reflection.

THE 180-DEGREE RULE

Mirrors have been well utilised throughout the history of cinema, in genres as diverse as film noir, fantasy, horror and art-house. The showpiece finale to *The Lady from Shanghai* (dir. Orson Welles, 1947) involving Welles himself, Rita Hayworth, and Everett Sloane chasing one another around a hall of mirrors is perhaps the most celebrated – and much imitated – example. As well as providing a device for creating tension and confusion in a dramatic showdown, mirrors are frequently used as a symbolic motif for character duality or secrecy, and in the fantasy/horror genres as a portal to other dimensions or the underworld.

However, there is a different cinematic convention that is worth reviewing in relation to left-right image reversal. When filming a dialogue scene between two characters directors typically follow a convention known as "the 180-degree rule". In a scene where two characters are having a face-to-face conversation, the rule determines that the action is filmed only from one side of the room, split into halves by an imaginary line drawn between the two characters. This means that if you see character one start the conversation from the viewpoint of character two this will be shot looking over the right shoulder of the character listening, but when we cut to the point of view of character one we will be looking over their left shoulder. Alternatively, we could have started looking over character two's left shoulder but to obey the rule, when we cut to character one's point of view we would be looking over their right shoulder.

Most of us are so accustomed to this convention whether in films or TV that we are mostly unaware of it (though if you were unfamiliar with this rule until reading this you may have fun checking whether directors stick to it or not when you next sit down to watch a movie or box set). However, when the rule is broken it becomes noticeable and unsettling. A direct and deliberate breaking of the rule features in Ava DuVernay's *Middle of Nowhere* (2012). Ruby (played by Emayatzy Corinealdi) is visiting her husband, Derek (Omari Hardwick) in prison and their conversation is followed in a series of typical over-the-shoulder shots which obey the 180-degree rule until a pivotal moment in the scene. At this point instead of seeing Ruby's reaction from behind Derek's left shoulder we switch to seeing her from over his right shoulder, and in confounding our expectations, it feels as if the rug has been pulled from underneath us.

Some conscious rule-breaking is also employed to good effect via a different mechanism in Nicolas Winding Refn's horror-thriller *The Neon Demon* (2016). In one of the first scenes in the movie, Jessie (played by Elle Fanning) and Ruby (played by Jena Malone) have a conversation while facing directly away from one another, but maintaining eye contact via their respective make-up mirrors. Ruby is doing most of the talking and we see her closer to us as she alternates her gaze between her own reflection as she applies her make-up and the reflection of Jessie. However, the optics of the scene mean that the 180 degree is broken and when Ruby looks away from her own reflection to that of Jessie, to us the viewers it looks as though she is looking the wrong way. Much of the action in this film is seen through mirrors of different sorts, and in a satirical horror movie extrapolating the worst excesses of the fashion/glamour industry the visual motif of the mirror as a reminder of both vanity and misleading appearances is not entirely subtle but does serve as a highly striking visual device.

SELFIES AND SELF-PORTRAITS

I had a moment of disorientation similar to that of breaking the 180-degree rule when taking a "selfie" with a group of friends in the pre-lockdown era. I utilised the timer mode as we couldn't fit everyone in if the shot was taken at arm's length, but the five- or ten-second delay gave a good opportunity to memorise what the final photo would look like as viewed on the phone's display screen. I was therefore a little surprised to find the image that was stored on my phone's memory was an inverted or "mirror image" of what I was expecting to see with the friend I thought would be on the left side on the right and vice versa. My friends were suitably unimpressed by my inability to grasp the concept of how selfies work, but in my defence it seems there is variation from one mobile phone manufacturer to another as to whether photos taken using the camera on the screen side of the phone are displayed "as taken" or in a mirror-image format.

My confusion prompts the question of whether there is a difference between a "selfie" and a self-portrait. The first photographic selfie is frequently attributed to Robert Cornelius, an amateur chemist and early photographer, who took a daguerre-otype exposure of himself in 1839, shown in Figure 4.4. Given that there is no arm stretching and that self-timing devices were not available at this point I had assumed on first seeing this image that Cornelius had used a mirror to achieve this image, perhaps positioned out of frame below his chest. However, this was not the case – Cornelius had set up the camera behind his family's shop in Philadelphia, before removing the lens cover, running into position in front of the camera, sitting in position for a minute, then rapidly returning to the camera to cover the lens again.[30] So, although requiring a bit more effort, this image is very much a selfie in the modern sense. In contrast to modern selfies, Cornelius is not looking directly at the camera lens, but seems more interested in something off to the side.

Most self-portraits through history have been performed using a mirror, resulting in left-right reversal. Contributing to the confusion in recent years is the emergence of the "mirror-selfie", and the "faked mirror-selfie", the first being a self-portrait using the main camera on a phone and a mirror, the second (a little perversely) mimicking this appearance but taken using the self-timer function on a camera or another individual as photographer.[31] Quite where my own contribution to the mirror-selfie genre (Figure 4.5) belongs in the classification is anyone's guess. A selection of famous self-portraits from simpler times are illustrated in Figures 4.6, 4.7, 4.8, and 4.9, showing Rembrandt, Dürer, Van Gogh, and Emile Bernard, respec-tively. While there are clear differences in style, the requirements of positioning a mirror impose certain constraints in terms of composition. Over the years, artists have produced some imaginative variations of the typical portrait aspect, looking towards the viewer type configuration, but this theme, dictated by the relative dimen-sions and alignment of the mirror, prevails amongst most hand-drawn or painted

FIGURE 4.4 Daguerreotype photograph of Robert Cornelius, 1839, frequently attributed to be the first "selfie". Public Domain Mark

FIGURE 4.5 Selfie (mirror-selfie?) photograph of the author taken within a mirrored lift, 2019. Image provided by the author.

FIGURE 4.6 Self-portrait of Rembrandt, Etching at a Window, by Rembrandt, 1648, Dutch print, etching on paper. Forty-two-year-old Rembrandt drawing on a copper-plate from his reflection in a mirror. Image credit: Everett Collection/ Shutterstock

FIGURE 4.7 Portrait of Albrecht Dürer, by Wenceslaus Hollar, 1645, Belgian print, etching on paper. Copied from original self-portrait of 1498 by Durer at age 26. Image credit: Everett Collection/ Shutterstock

FIGURE 4.8 Self-portrait, by Vincent van Gogh, 1889, Dutch Post-Impressionist painting, oil on canvas. Image credit: Everett Collection/ Shutterstock

FIGURE 4.9 Self-portrait, by Emile Bernard, 1897, French painting, oil on canvas. Image credit: Everett Collection/ Shutterstock

self-portraits. We will consider the confines of a rectangular image format further in Chapter 5, but this specific constraint is also of particular relevance to the construction of world maps.

GLOBAL MAPS

A short walk from where I live in Edinburgh is the Grange Cemetery, within which are buried a number of notable historic figures of Scottish life including Rev Thomas Chalmers (founder of the Free Church of Scotland), Charles MacLaren (founder of the Scotsman newspaper), and Canon Edward Hannan (founder of Hibernian Football Club).[32] One burial that is currently omitted from the visitor information board at the gates to the cemetery is that of James Gall, a minister of the Free Church of Scotland who established the Carrubbers Close mission. Rev Gall was something of a polymath being an author, publisher, sculptor, astronomer and, of most relevance to this chapter, a cartographer. He had acquired some knowledge of map-making while working in his father's printmaking firm (which specialised in maps) during his early adult life, and he combined this with his understanding of astronomy to formulate what is known as the Gall orthographic projection. This is a mathematical description of how to project topographical coordinates from a spherical structure (such as the Earth) onto a flat, planar surface (such as a rectangular map).[33]

Gall presented this projection, along with two other systems (stereographic and isographic projections) at a scientific meeting in Glasgow in 1855, eventually publishing a description in 1885. This did not catch on in a big way during Gall's lifetime, and towards the end of his life, he is said to have reflected that the only person to have made use of it was himself. This may have been related to his own reticence – in relation to the orthographic projection he acknowledged that "the geographical features are more distorted on this than on any of the others" and suggested that of the three world maps he had proposed the stereographic was the best.[34] Although his cartographic endeavour was largely overlooked in his lifetime, his other numerous achievements, particularly in social welfare, ensured attendance of over 600 at his funeral in 1895.[33]

Many years later in 1973, the German historian Arno Peters summoned some 350 international journalists to a press conference in Bonn (capital of the Federal Republic of West Germany as it was), where he announced a new world map based on the "Peters projection". It did not take other cartographers long to notice that this was all but identical to what Gall had described.[35] However, while Gall was rather reticent in promoting his projection, Peters proved to be quite the salesman. What became known as the "Gall-Peters projection" has since become adopted by a number of international organisations including UNESCO as their preferred method of depicting the world in rectangular, flat format.[36] In 2017 Boston public schools became the most recent example of a high-profile institution switching their world maps to this format.[37]

The Gall-Peters projection has been advocated as an improvement over the traditional Mercator projection which had dominated Western world maps for 500 years since its introduction by the Flemish geographer and cartographer Geradus Mercator in 1569.[38] In the Mercator projection (originally designed to assist nautical navigation by enabling key shipping routes to be plotted as straight lines) land masses become increasingly inflated in size the further away from the equator they are, whereas in the Gall-Peters projection effectively the reverse is the case. This is well illustrated by looking at the relative sizes of Greenland and Africa in Figure 4.10, a world map rendered using the Gall-Peters projection, and Figure 4.11, a world map based on Mercator projection proportions. Advocates of the Gall-Peters projection would argue that given that the size of Africa is, in reality, 14 times that of Greenland, this format provides a more accurate representation of the world. Furthermore, the prevalence of the Mercator world map in geography books, atlases, and school rooms was characterised as "cartographic imperialism" by Peters.[38] The political dimensions of these differing world maps was highlighted in an episode of the TV show *The West Wing*.[39]

However, while the Gall-Peters projection has attracted support for the adjusted size scale of Africa and South America, Peters himself was a polarising personality. The proposal that the Mercator version of the world was a significant distortion was accepted by most cartographers, but his assertion that his projection was the solution was not. As Carl Friedric Gauss had proven in the 1820s, it is impossible to accurately project a sphere onto a plane surface without introducing some degree of distortion.[40] Critics of Peters also suggested that he oversold the extent to which the Mercator projection had dominated the field. It seems likely that had he proposed his one-man solution to centuries of cartographic oppression of developing countries in more recent times he would have been accused of a "white saviour complex". Following a period of controversy and division amongst the cartographic community during the 1970s and 1980s several North American geographic organisations composed a resolution which rejected *all* rectangular world maps, pointing out the severe distortions and conceptual errors that such representations promote.[41]

This "plague on both your houses" approach seems to have had limited success in eliminating rectangular world maps so far, and the controversy this episode generated has yet to fully subside almost half a century later. A less acrimonious debate occurred more recently in the UK in response to the BBC weather map. After a few years using a tilted or curved map of the UK for TV weather forecasts the BBC returned to a flat map format in 2018, prompting a social media response welcoming the "return of Scotland" as the northernmost parts of the country resumed the larger size they had enjoyed prior to the use of the curved map.[42] On this occasion the modification was near universally approved of, although the increased prominence of Scotland in the UK landscape was seized upon by supporters of Scottish independence, underlining the inescapable political dimensions of map-making.

This may all seem far removed from the business of looking at medical images, but take a look at Figures 4.12 and 4.13. Figure 4.12 shows an MRI of the brain in which the data has been manipulated, such that the curved surface of the brain has been "rolled flat" to make it easier to interpret the cortical architecture. This is directly analogous to the process of flattening out the surface of the world to produce a rectangular map. Some similar flattening out is shown in Figure 4.13, again an MRI of the brain, but this time rendered using white matter tractography. While the degree of distortion is relatively modest, the same process of tailoring the images to a specific purpose is very much at play.

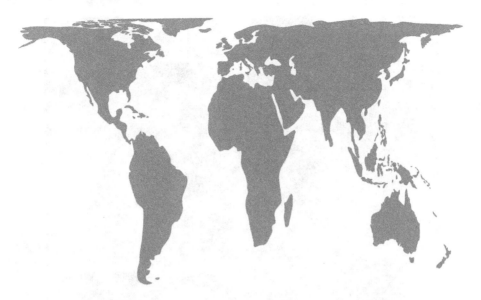

FIGURE 4.10   A world map rendered using the Gall-Peters projection. Image credit: Jason_ Li/ Shutterstock

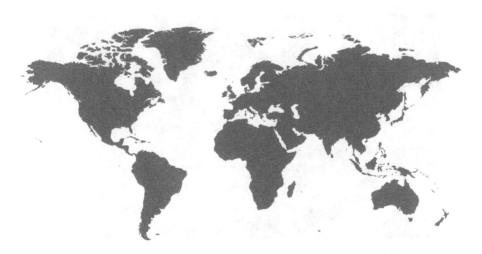

FIGURE 4.11   World map based on Mercator projection. Image credit: timyee/ Shutterstock

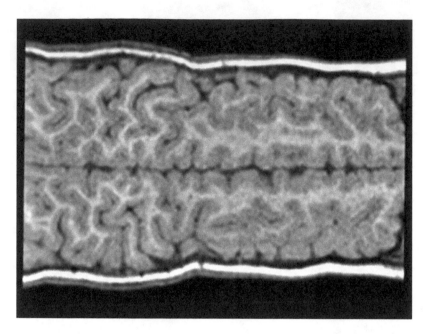

FIGURE 4.12 MRI of the brain in which the data has been manipulated, such that the curved surface of the brain has been "rolled flat" to make it easier to interpret the cortical architecture. This is directly analogous to the process of flattening out the surface of the world to produce a rectangular map. Image provided by the author.

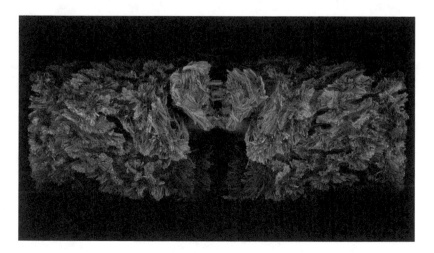

FIGURE 4.13 MRI of the brain rendered using white matter tractography. The image shows a Mercator transformation of the 3D space embedding a dense human brain tractography. The cerebellum is at the top of the image at the midline, immediately to the sides are the left and right temporal lobes, and the interhemispheric margin appears at the left and right borders of the image. Credit: Unfolded brain, MRI. Katja Heuer and Roberto Toro. Wellcome Collection Attribution 4.0 International (CC BY 4.0).

TUBE MAPS AND ROADMAPS

There are numerous textbooks titled Radiographic Atlas of (something or other) and plenty of anatomy atlases. While at medical school in London one anatomy textbook caught my eye as a striking example of the overlap between the design aesthetics of modern maps and road atlases, and the representational considerations of depicting human anatomy. Admittedly this book did not call itself an atlas (it was *Instant Anatomy* by Robert H. Whitaker)[43] but the diagrams of major blood vessels and their tributaries and those of major nerves and their branches reminded me of the Underground tube map that I was seeing on a daily basis at the time. Take a look at Figures 4.14 (an illustration of the brachial plexus from *Instant Anatomy*) and 4.15 (the London Underground map) to make your own comparison. The Tube map was devised in 1933 by Harry Beck, who was working as an electrical draughtsman for the Underground at the time. In applying the schematic approach of the circuit diagrams he was drawing in his day job, Beck's tube map distorts the distances and spatial relationships between stations.[44] However, in losing geographical or topographical accuracy it gains clarity and comprehensibility. By using a similar approach in anatomical diagrams the risk is that the drive for clarity compromises the accuracy of the structures they claim to represent. In the case of *Instant Anatomy*, the concise text and schematic-style diagrams were explicitly intended as a revision tool to summarise key points, but the blurring of stylistic and scientific choices can sometimes be less straightforward.

In radiographic practice, there are certainly situations where the nature of the images seems to be more tailored to "anatomy as it is useful to see it" rather than "anatomy as it is". At the start of interventional radiology procedures such as placing a stent across a narrowed artery or coiling an aneurysm, X-ray contrast media is injected into blood vessels to produce angiographic images which serve as a guide for the radiologist to introduce a wire to the site of interest. These baseline images acquired at the beginning are known as "the roadmap", and depending on where the anatomical location of the problem is, the radiologist may need to negotiate the wire through multiple turns based on this map, hopefully not getting lost on the way. The analogy of the map in this situation is readily understandable (although you could argue that tubular blood vessels are more akin to the tunnels of the Underground than flat roads). The roadmap images often employ a technique called digital subtraction angiography ("DSA"), which in simple terms gets rid of unnecessary clutter in the images such as bones or bowel gas. The technique produces images in which the blood vessels of interest are shown in dark black on a (mostly) featureless white (or perhaps pale grey) background, as shown in Figure 4.16.

Here, our blood vessels have been imaged in a fashion very similar to anatomical figures we will encounter in Chapter 6 and employing the kind of emphatic black lines that Daumier and Manet would approve of (discussed in Chapter 3). However, the vessels are not outlined, but actually *are* lines themselves, taking on the symbol-like appearance of roads, train tracks, or tube routes on a map. So, while the term "roadmap" may be an analogy as no vehicles are being driven, it is difficult to argue that these images are not – in a very real sense – a map. It is also incontestable that to produce this near-two tone, stylised image, there has been substantial manipulation of the images. I have also further manipulated the image by displaying

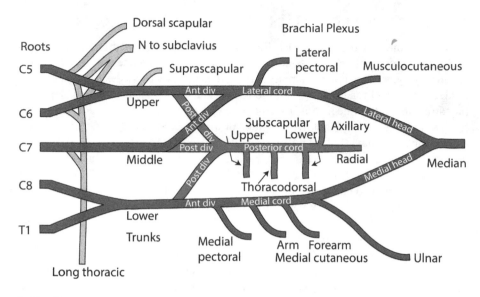

FIGURES 4.14 Illustration of the brachial plexus from *Instant Anatomy*. Image courtesy of Robert Whitaker / instantanatomy.com.

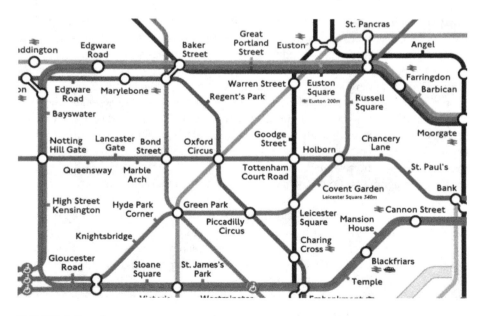

FIGURE 4.15 Cropped segment of the London Underground map. From a photograph by Thinglass/ Shutterstock

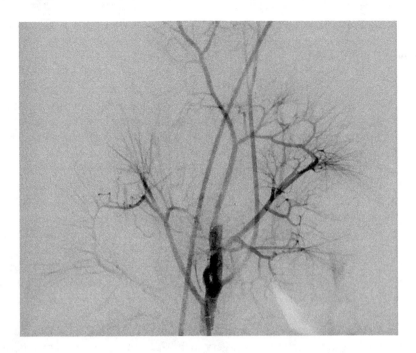

FIGURE 4.16   Digital subtraction angiography image of abdominal blood vessels (inverted). Image courtesy of Dr G Weir.

it "upside down", so that as well as resembling a map, it is similar in appearance to the Japanese watercolour painting in the previous chapter (Figure 3.3).

For another example of image manipulation, return to Chapter 2 and take a look at Figures 2.2 and 2.3. In Figure 2.3 both the fovea and the optic disc are illustrated at the back of the eye, shown in a cross-sectional diagram in a sagittal plane. This is an anatomical distortion as both these structures could never be seen on the same slice in this orientation (the two structures are shown on top and below, when the reality is that they are side by side). That, you may think, is fair enough – it is similar to the type of diagrammatic free license we saw in *Instant Anatomy*. Surely Figure 2.2, however, should be showing the anatomical truth? It is a scientific image not a graphically designed diagram, after all. But again, we can see both the fovea and the optic disc, in a very similar configuration to that seen in the diagram of Figure 2.3. While this image is produced using a scientifically robust technique, there has been some sleight of hand in how the composite image has been arranged to deliver appearances which meet the expectations of the viewer.

BRAIN ATLASES

The comparison of medical images to maps is unavoidable in relation to MRI brain atlases (also known as brain templates). These have proliferated in recent decades with a wide range of research applications and a growing number of clinical uses.

Such atlases are generated by combining MRI scan data from a number of different individuals to produce an "average" representation of the brain. This average brain can then be utilised as a comparator to an individual's (or group of individuals) brain scan data, or to help generate algorithms to digitally segment the brain into particular regions of interest. By using a standardised set of coordinates to locate particular brain regions the ambition is to enable more accurate localisation of disease in individuals and to enable insights into normal brain functioning more broadly.[45]

The difficulty of this undertaking is that there is significant variation in the shape of the brain from one individual to another, and within any one individual over the course of their life. To compensate for this some atlases have combined scan data from hundreds and sometimes thousands of individuals in an attempt to average out this variability. The problem is that combining multiple brains of slightly different shapes leads to fuzziness of the margins between different brain structures. Many templates acknowledge the uncertainty generated by this process, labelling themselves "probabilistic" atlases. Like locating an electron in an atom, we can only determine the location of particular brain structures to a level of probability rather than absolute certainty. The more robustly an atlas is constructed, the higher the likelihood of anatomical accuracy. Conversely, deficits in the design of the atlas reduce the probability of accuracy.

Many MRI atlases are in some way derived from Talairach Coordinates devised as a tool to assist stereotactic neurosurgery in 1967,[46] and the subsequent Talairach-Tournoux Atlas of 1988,[47] based on the dissection of a solitary brain. As with the various world maps we have considered, Talairach had to decide what axes and landmarks were to be used to base the coordinate system on. He selected two white matter tracts that cross the midline of the brain in front of and behind the thalamus, known as the anterior and posterior commissures, as the key landmarks. These are located quite centrally within the brain, reflecting his interest as a neurosurgeon in accurately locating deep brain structures, rather than more readily accessible areas at the cortical surface. The original coordinate system also omitted the cerebellum (a brain structure smaller in size than the cerebral cortex, but actually containing a higher number of neurons). Utilising the Talairach coordinate system as a template to configure MRI datasets posed numerous challenges. A combination of the coordinates being based on the anatomy of a single individual and the choice of such centrally located anchoring points meant that structures located more peripherally, particularly areas of the undulating cortical surface, were frequently mismatched by a number of millimetres.[48] In these instances our first bilateral ancestor *Ikaria wariootia* could have happily fitted into the space available between the actual location of interest in the brain and where the atlas indicated it to be. While more recent atlases have either substantially modified the Talairach system or done away with it altogether, a survey conducted in 2016 found that some 20% of MRI brain atlases were derived in some fashion from Talairach and Tournoux.[45]

There is also a compromise to be achieved in relation to how representative the atlas is to the individual or population group being studied. It has been established, for example, that using an adult brain atlas to analyse children's brain imaging data causes misclassification of tissues and structures.[49] Likewise, using an atlas composed of young adult brains causes warping of the mapping process when applied to older patients in whom brain volume is typically smaller.[50] In theory, we would want the closest match possible so the brains from which the atlas is composed should come from individuals of similar age, gender, and ethnicity. There is also an argument for matching medical conditions such as hypertension or diabetes and environmental factors such as smoking or alcohol consumption.[45] In addition, research has shown London taxi drivers to have slightly different shaped brains compared to the general population – their hippocampi (brain structures concerned with memory) are a little larger, thought to relate to having had to memorise the entire road system of London.[51] So perhaps atlases also need to take occupation into account. Taking this close-matching approach to an extreme, the only reliable atlas would be one based solely on imaging data from the individual concerned, but of course this would be self-defeating as we lose the ability to compare this with our expected version of normality, whatever that might be.

The logistics of acquiring good quality MRI scans suitable for use in these atlases has meant there is a potential skew in the age distribution of these atlases. Lying still for a number of minutes at a time in the unfamiliar environment of an MRI scanner, together with considerations such as getting to and from the scanning centre and being able to provide informed consent has meant atlases based on healthy, young adults tend to be better populated (or at least composed of better quality imaging data) than those of infants, children, and the elderly.[45] This is problematic as it is at the extremes of age where brain atlases are potentially of most clinical use – identifying a structural abnormality that may account for a young child not meeting their expected developmental milestones, or demonstrating reasons for a deterioration in cognitive function in an elderly individual. While that does not imply the healthy young adult atlases are a waste of time, it does serve to highlight an issue that will become of increasing importance as machine learning is utilised in clinical practice – the validity of the "learning population" in relation to the "clinical population".

THE FOURTH DIMENSION

We have considered some of the spatial dimensions of maps and images but these are often inseparable from time. Certainly within radiology, the date and time an examination has been performed are essential for its meaningful interpretation. In my own field looking at the imaging of children this may relate to whether the developmental stage of bones or other bodily structures is appropriate for their age, but in general the chronological order of imaging is crucial to make sense of a patient's clinical state. Knowing if the tumour is getting bigger or smaller, an infection getting worse or better, a device moved from the wrong position to the correct one or vice versa all relies on knowing when each respective imaging examination was acquired. Timing of any imaging examination is therefore of critical importance, whether it be the date ordering of serial examinations, or in relation to dynamic elements of an image acquisition such as injection of intravenous contrast in which timing needs to be accurate to a matter of seconds, sometimes split seconds.

Timing has always been an integral part of the process of navigation, from the approximate (sunset, sunrise, midday), the accurate (increasingly reliable clockwork ship chronometers developed during the 16th and 17th centuries), and the super-accurate capabilities of 21st-century timepieces. The GPS navigation system relies on 24 satellites orbiting the globe some 20,000 km above the earth's surface. To successfully triangulate the position of your mobile phone every satellite requires a synchronised time signature. Compared to clocks running at sea level, those in the satellites run 38 microseconds faster (they are sped up by 45 microseconds due to a weaker gravitational field, but slowed down by 7 microseconds due to travelling at speeds of around 14,000 km/hour). These effects, predicted by Einstein's theory of special relativity, are factored in to enable the GPS system to function.[52]

Speeding around the planet in a low earth orbit also caused a very specific navigational challenge to Malaysian astronaut Sheik Muszaphar Shukor during his stay on the International Space Station in 2007. As a devout Muslim, Shukor sought advice on what direction he should consider the *qibla* to be in this highly dynamic situation. Malaysia's National Fatwa Council gave the very pragmatic advice that he should assign the *qibla* "based on what is possible", reflecting the broader view that it is the intention, rather than precise geographical accuracy that is of importance. In this situation, the change in direction, as described by more traditional criteria, could have been of a head-spinning rate, altering by 180 degrees during the length of a single prayer.[53]

The *qibla* was in fact originally aligned towards Jerusalem, being changed to the direction of Mecca in 624 CE.[54] So you might say very little is set in stone. Indeed, even things that *are* set in stone – specifically the magnetic field alignment of the Earth as recorded in volcanic or sedimentary rocks – are subject to change. Ferromagnetic minerals found within such stones have demonstrated some 183 reversals of the direction of the Earth's magnetic field over the past 83 million years.[55] The most recent reversal occurred about 780,000 years ago (known as the Brunhes-Matuyama reversal), and the pattern in recent geological history is of a switch approximately every 200,000 to 300,000 years, suggesting we are due for a reversal before too long.[56] Such reversals are thought to occur over a period of hundreds or thousands of years during which magnetic polarity is in a chaotic state of flux, and a conventional compass – such as that radiographed by Röntgen 130 years ago – would be of little use for navigational purposes. The relentless march of entropy means it is an irrefutable necessity that all maps, whether of geographic terrain or of the human body, are "of their time". It is also the case that in requiring a specific frame of reference to be correctly interpreted, both geographic maps and imaging of the body require a very particular point of view, a theme we will explore further in the next chapter.

REFERENCES

1. Jerry Brotton, *History of the World in Twelve Maps*, Allen Lane, London, 2012.
2. Ibid p 57

3. Richard Brown, The east, time, eternity, the universe and the origin of all things, Ancient Hebrew Research Centre, https://www.ancient-hebrew.org/philosophy/east-time-eternity-the-universe-and-the-origin-of-all-things.htm#, last accessed 17/07/2021.
4. Christopher Heffner, Out of the question, *Church Times*, https://www.churchtimes.co.uk/articles/2008/11-january/features/features/out-of-the-question, last accessed 17/07/2021.
5. S. Heblich, A. Trew and Y. Zylberberg, *East Side Story: Historical Pollution and Persistent Neighborhood Sorting*, November 08, 2016. CESifo Working Paper Series No. 6166, https://ssrn.com/abstract=2884598, last accessed 17/07/2021.
6. A. Jerry Brotton, *History of the World in Twelve Maps*, Allen Lane, London, 2012, p 57–58.
7. Ibid p 140
8. Jasper Burford and Ellen Burford, Mandarin Chinese to English dictionary—Using Pinyin, pp 13–14, http://www.jaspell.uk/pinyin_introductory/pinyin_dictionary_m_to_e_a5c1_20170909.pdf, last accessed 17/07/2021.
9. Paul Anthony Jones, 5 essential types of words that some languages do without, *Mental Floss*, https://www.mentalfloss.com/article/80947/5-essential-types-words-some-languages-do-without#:~:text=Guugu%20Yimithirr%2C%20or%20Guguyimidjir%2C%20has,%2C%20south%2C%20east%20and%20west, last accessed 17/07/2021.
10. J. B. Haviland, Guugu Yimithirr cardinal directions, *Ethos* 26(1), March 1998, 25–47. https://pages.ucsd.edu/~jhaviland/Publications/ETHOSw.Diags.pdf, last accessed 17/07/2021.
11. S. D. Evans, I. V. Hughes, J. G. Gehling and M. L. Droser, Discovery of the oldest bilaterian from the Ediacaran of South Australia, *Proc. Natl. Acad. Sci. USA* 117(14), April 2020, 7845–7850. doi: 10.1073/pnas.2001045117
12. Wikipedia, Fiddler crab, https://en.wikipedia.org/wiki/Fiddler_crab, last accessed 17/07/2021.
13. F. Raffini and A. Meyer, A comprehensive overview of the developmental basis and adaptive significance of a textbook polymorphism: Head asymmetry in the cichlid fish Perissodus microlepis, *Hydrobiologia* 832, 2019, 65–84. doi:10.1007/s10750-018-3800-z
14. P. Danaisawadi, T. Asami, H. Ota et al. A snail-eating snake recognizes prey handedness, *Sci Rep* 6, 2016, 23832. doi:10.1038/srep23832
15. Wikipedia, Jeremy the snail, https://en.wikipedia.org/wiki/Jeremy_(snail), last accessed 17/07/2021.
16. A. Catana and A. P. Apostu, The determination factors of left-right asymmetry disorders- a short review, *Clujul Med.* 90(2), 2017, 139–146. doi:10.15386/cjmed-701
17. C. V. E. Wright, Mechanisms of left-right asymmetry: What's right and what's left?, *Developmental Cell* 1(2), 2001, 179–186, ISSN 1534-5807, doi:10.1016/S1534-5807(01)00036-3.
18. P. Pennekamp, T. Menchen, B. Dworniczak and H. Hamada, Situs inversus and ciliary abnormalities: 20 years later, what is the connection?, *Cilia* 4(1), 2015, 1. Published 2015 Jan 14. doi:10.1186/s13630-014-0010-9
19. H. R. Haththotuwa and S. W. Dubrey, A heart on the right can be more complex than it first appears, *BMJ Case Rep.* 2013. bcr2013201046. Published 2013 Sep 19. doi:10.1136/bcr-2013-201046
20. Anatomy Notes, http://anatomynotes.blogspot.com/2006/03/backward-chest-x-ray-in-scrubs.html, last accessed 17/07/2021.
21. M. Fuller, Side marker creep: have radiographers changed their sidemarker habits?, *J Med Radiat Sci* 63, 2016, 143–144. doi: 10.1002/jmrs.181https://onlinelibrary.wiley.com/doi/pdf/10.1002/jmrs.181
22. Behold AI, https://behold.ai/how-it-works/, last accessed 17/07/2021.

23. E.H. Gombrich, *Art and Illusion: A Study in the Psychology of Pictorial Representation*. 5th edition, Phaidon Press, London, 1977, p 33.
24. C.-C. Carbon and A. Pastukhov, Reliable top-left light convention starts with early renaissance: An extensive approach comprising 10k artworks, *Front. Psychol.* 9, 2018, p 454. https://www.frontiersin.org/article/10.3389/fpsyg.2018.00454 doi:10.3389/fpsyg.2018.00454
25. P. Lanthony, Les peintres gauchers, *Rev. Neurol. (Paris)* 151, 1995, 165–170.
26. https://artintheblood.typepad.com/arthistcert2012/2012/11/week-7-reading-left-and-right-in-art-history.html
27. J. Hall, *The Sinister Side: How Left-Right Symbolism Shaped Western Art*, Oxford University Press, Oxford, 2008.
28. Ibid, p 17
29. Ibid, p 18
30. Adam Green (ed.), Robert Cornelius' self-portrait: The first ever "Selfie" (1839), *Public Domain Review*, https://publicdomainreview.org/collection/robert-cornelius-self-portrait-the-first-ever-selfie-1839, last accessed 17/07/2021.
31. Ellie Violet Bramley, 'The fakery is all part of the fun': The hoax of the mirror selfie, *Guardian*, 22/03/2021, https://www.theguardian.com/fashion/2021/mar/22/the-fakery-is-all-part-of-the-fun-the-hoax-of-the-mirror-selfie, last accessed 17/07/2021.
32. Grange Cemetery: Some notable burials. Grange Association Edinburgh, http://gaedin.co.uk/wp/new-history/cemetery, last accessed 17/07/2021.
33. Wikipedia, James Gall, https://en.wikipedia.org/wiki/James_Gall, last accessed 17/07/2021.
34. J. Brotton, *A History of the World in Twelve Maps*, Allen Lane, London, 2012, p 395.
35. Ibid p 378.
36. Wikipedia, https://en.wikipedia.org/wiki/Gall%E2%80%93Peters_projection, last accessed 17/07/2021.
37. Joanna Walters, Boston public schools map switch aims to amend 500 years of distortion, *The Guardian,* 23/03/2017, https://www.theguardian.com/education/2017/mar/19/boston-public-schools-world-map-mercator-peters-projection, last accessed 17/07/2021.
38. Wikipedia, Mercator projection, https://en.wikipedia.org/wiki/Mercator_projection#:~:text=The%20Mercator%20projection%20(%2Fm%C9%99r,cartographer%20Gerardus%20Mercator%20in%201569, last accessed 17/07/2021.
39. "The West Wing" Season 2 Episode 16 https://youtu.be/vVX-PrBRtTY, last accessed 17/07/2021.
40. J. Brotton, *A History of the World in Twelve Maps*, Allen Lane, London, 2012, p 12.
41. A. H. Robinson, Rectangular world maps—No!, *Prof. Geogr.*, 42(1), 1990, 101–104, doi:10.1111/j.0033-0124.1990.00101.x
42. Hamish Mackay, BBC weather redesign - Viewers hail 'Scotland's return', *BBC News*, 06/02/2018, https://www.bbc.co.uk/news/uk-42945763, last accessed 17/07/2021.
43. Robert H. Whitaker and Neil R. Borley, *Instant Anatomy*. Wiley–Blackwell, London, 2005.
44. Harry Beck's tube map, Transport for London, https://tfl.gov.uk/corporate/about-tfl/culture-and-heritage/art-and-design/harry-becks-tube-map, last accessed 17/07/2021.
45. D. D. Alexander, D. Shenkin Susan, A. Devasuda, L. Juyoung, B. C. Manuel, R. David, P. Boardman James, W. Adam, E. Job Dominic and M. Wardlaw Joanna, Whole brain magnetic resonance image atlases: A systematic review of existing atlases and caveats for use in population imaging, *Front. Neuroinform.* 11, 2017, https://www.frontiersin.org/article/10.3389/fninf.2017.00001

46. J. Talairach, G. Szikla, P. Tournoux, A. Prosalentis, M. Bordas-Ferrier, L. Covello, et al. *Atlas D'Anatomie Stereotaxique du Telencephale*, Masson, Paris, 1967.

47. J. Talairach and P. Tournoux, *Co-planar Sterotactic Atlas of the Human Brain: 3-dimensional Proportional System: An Approach to Cerebral Imaging*, Georg Thieme Verlag, Stuttgart, 1988.

48. A. R. Laird, J. L. Robinson, K. M. McMillan, et al. Comparison of the disparity between Talairach and MNI coordinates in functional neuroimaging data: validation of the Lancaster transform, *Neuroimage* 51(2), 2010, 677–683. doi:10.1016/j.neuroimage.2010.02.048

49. U. Yoon, V. S. Fonov, D. Perusse and A. C. Evans, The effect of template choice on morphometric analysis of pediatric brain data, *Neuroimage* 45, 2009, 769–777. doi:10.1016/j.neuroimage.2008.12.046

50. R. L. Buckner, D. Head, J. Parker, A. F. Fotenos, D. Marcus, J. C. Morris, et al. A unified approach for morphometric and functional data analysis in young, old, and demented adults using automated atlas-based head size normalization: reliability and validation against manual measurement of total intracranial volume, *Neuroimage* 23, 2004, 724–738. doi: 10.1016/j.neuroimage.2004.06.018

51. E. A. Maguire, D. G. Gadian, I. S. Johnsrude, C. D. Good, J. Ashburner, R. S. Frackowiak and C. D. Frith, Navigation-related structural change in the hippocampi of taxi drivers, *Proc. Natl Acad. Sci. USA.* 97, 2000, 4398–4403. doi:10.1073/pnas.070039597

52. B. Cox and J. Forshaw, *Why Does E=mc2: (and Why Should We Care?)*, Da Capo Press, Boston, 2009, p 235.

53. A Muslim astronaut's Dilemma: How to face Mecca from space, Wired, https://www.wired.com/2007/09/mecca-in-orbit/#:~:text=From%20ISS%2C%20orbiting%20220%20miles,course%20of%20a%20single%20prayer, last accessed 17/07/2021.

54. Wikipedia, Qibla, https://en.wikipedia.org/wiki/Qibla, last accessed 17/07/2021.

55. Wikipedia, Geomagnetic reversal, https://en.wikipedia.org/wiki/Geomagnetic_reversal, last accessed 17/07/2021.

56. 2012: Magnetic pole reversal happens all the (geologic) time, NASA, https://www.nasa.gov/topics/earth/features/2012-poleReversal.html, last accessed 17/07/2021.

5 Point of View

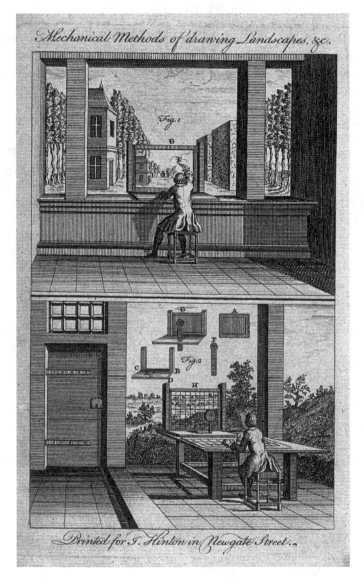

A seated man drawing a view onto a framed piece of glass and a seated man drawing from a perspective grid. Etching. Credit: Wellcome Collection. Attribution 4.0 International (CC BY 4.0).

DOI: 10.1201/9780367855567-5

As we saw in the previous chapter, images of all sorts can be influenced by political and cultural factors. As well as the image itself, the viewpoint of the individual looking at the image is also crucial. Both "point of view" and "viewpoint" are often used in a political context, as are "worldview", "outlook" and in recent years "optics" has also acquired a meaning related to how appearances and events can be manipulated for political or other ends. Both *The Observer* and *The Spectator* keep an eye on political events but from very different perspectives. Since the 18th century the word "perspective" itself has been understood to mean a way of thinking about something, as much as a way of seeing something.[1] We shall consider some examples of medical images employed to convey overt political messages in Chapter 7. In this chapter, the political dimension is less concerned with the business of governing countries, and more related to the art of managing limited resources, though we shall also see that it can be difficult to fully remove political aspects from imaging the body.

Following on from our exploration of how images may become distorted in relation to how they are constructed in the previous chapter, this chapter initially considers linear perspective. We shall see how this technique perpetuates two problematic aspects of cartography – a rectangular format of the image, and the requirement for a very specific, unique point of view. Expanding from this we shall see that at the point of their interpretation *all* images, regardless of the viewpoint from which they were taken, or the mechanisms used in their construction (including those of modern cross-sectional imaging), require the specific point of view of the individual doing the interpretation.

WINDOWS ON THE WORLD, WINDOWS ON THE BODY

In an excellent paper on the links between the use of a grid system in delivering geometric perspective in artwork and its origins within cartography, Barry Smith emphasises the early renaissance belief that the work of the artist should represent the visible world as if the observer of the picture were looking through a window.[2] As a radiologist I frequently use the terms "lung windows" or "bone windows" to refer to the specific range of contrast values applied to particular densities on CT images to optimise the display of particular structures or tissues (discussed in Chapter 6). However, prior to embarking on radiology I remember getting confused by these terms, thinking that "lung windows" simply referred to the images (or "slices") in which the lungs were included, the window label indicating the rectangular frame within which the lungs were displayed ("Microsoft Windows" likely to have contributed to my confusion). Windows allow us to see through walls in the same way that medical imaging allows us to see through the body. In ultrasound we also refer to "acoustic windows" – particular positions and angles to place the transducer to get the best visualisation of organs, avoiding impenetrable structures such as ribs or gas-filled bowel loops. In this context, it is noteworthy that the word *perspective* is derived from the Latin *perspicere*, in turn originating from the Proto-Indo-European roots *per* (meaning "through") and *spek* (meaning "to observe, look at").[1] Windows allow us to look beyond otherwise impenetrable structures, and perspective, as originally derived, means "to see through".

In Chapter 1 we briefly considered the "reticolato" devised by Leon Battista Alberti, and utilised by Leonardo da Vinci, as an early precursor of digital image capture technology. In pioneering the technique of linear perspective, Alberti is known to have been inspired by recent advances in optics, but is also thought to have been heavily influenced by Ptolemy's *Geographia*, a work dating from around 140 CE, which re-emerged in Florence in 1400.[2] In this work Ptolemy describes the possibility of using a regular mathematical grid system to map the entire known world, very similar in methodology to the Gall-Peters projection discussed in the previous chapter.[3] We have examined the difficulties created by flattening out the spherical globe into a two-dimensional rectangular image in relation to cartography and the same challenges – albeit on a different scale – apply to any image rendered using the technique of linear perspective.

While portholes of ships and lancet panes in churches provide some exceptions, the rectangular dimensions of bricks have ensured that the large majority of window panes are similarly rectangular. Likewise, the combination of the rectangular dimensions of the reticolato and the convenience of rectangular canvas frames helped to establish the rectangle as the conventional format for two-dimensional representational art, emulating a "window-like" appearance in the process. The notation of "portrait" and "landscape" for the vertical or horizontal alignment of the rectangle respectively has passed from oil paintings to paper sizing, Powerpoint slides, and the orientation of computer monitors to name just a few. In modern life the rectangular image format is all-pervasive. I type this text onto a rectangular laptop screen whilst periodically looking out of the rectangular window onto the street. As I do so, a text arrives on my phone and … well, you get the picture.

So, the self-conscious intent of renaissance era artists to replicate the experience of looking out of a window in the use of linear perspective contributed to a rectangular vision of the world, with all the associated constraints. In addition, numerous art critics and historians have highlighted the inherent limitation of linear perspective in that it shows the action from one very specific point of view. In *Ways of Seeing*, John Berger suggests that while perspective arranged the world in a fashion similar to how God was thought to survey the universe, there was a central contradiction in that any human observer can only ever be in one place at any time.[4]

Other critics argue that the inherently illusionistic images constructed using linear perspective are so ingrained within western visual traditions, that we are likely to interpret any rectangular image as though viewing the action through a window.[5]

ANAMORPHOSIS

The illusionistic nature of images constructed using linear perspective is well illustrated by anamorphic compositions. These are images which appear to be distorted when viewed from a frontal position, but which assume normal proportions when looked at from a particular viewpoint or angle. Perhaps the most famous example of such an image is the stretched looking skull in the foreground of *The Ambassadors* by Hans Holbein the Younger (1533), shown in Figure 5.1. The viewer is required to be almost level with the frame of the picture before the skull – rather magically – takes on a normal appearance (replicated in Figure 5.2). In this situation the viewer has to actively reposition themselves from their usual viewing position to get the benefit, usually resulting in a queue to have a turn at the National Gallery in London.

FIGURE 5.1   Jean de Dinteville and Georges de Selve ("The Ambassadors"). Hans HOLBEIN the Younger. © The National Gallery, London

FIGURE 5.2   Jean de Dinteville and Georges de Selve ("The Ambassadors"). Hans HOLBEIN the Younger. (cropped segment, stretched to mimic the effect of viewing the painting level with the frame). © The National Gallery, London

However, art has often utilised anamorphosis to allow the viewer to perceive correct proportions from the most typical viewpoint. Large-scale sculptures towering over people on the ground have been proportioned to look correct from a street-level vantage point through history, and even some cave paintings are said to have exploited anamorphic principles to enable the best visual experience from a particular location within the cave.[6]

In more recent times, most readers will be familiar with the "keystone correction" facility of digital projectors that allows the projected image to retain a rectangular shape, rather than trapezoidal shape that would otherwise result if the projector is at an angle to the surface being projected onto. Sports fans will certainly have come across adverts painted onto football and rugby pitches for major matches, in which the proportions are specifically tailored to the "overview" position of the TV cameras from which the majority of the action will be broadcast. In this situation, the adverts will either be upside down or severely distorted for the large majority of the thousands of spectators within the stadium itself, but correctly aligned for the millions of viewers at home.

Prior to digital radiography, a tip I was given when looking for a pneumothorax on a (hard copy) chest X-ray was to remove the radiograph from the viewing box and then angle the film such that it was almost horizontal. This anamorphic manipulation is said to improve the visibility of the faint line of the lung edge, although I cannot recall having spotted any pneumothoraces in the pre-digital era using this technique. The availability of contrast manipulation, zooming, and edge-enhancement on digital reporting stations has now made this niche manoeuvre largely redundant.

However, more broadly the principle of anamorphosis is of fundamental importance in imaging techniques. In plain film radiography, the quality of the image projected onto the X-ray film or detector plate depends on both the distance between the source of X-rays and the subject, and the distance between the subject and the detector plate. For the sharpest image the first of these should be as long as possible (such that the X-rays arrive at the subject more parallel than divergent), and the second of these as short as possible (to prevent magnification and blurring of structures). The principles are the same as those used when casting shadow puppets onto a wall, or tracing the profile of a lover's silhouette. While keeping the patient close to the detector plate is generally achievable, practical considerations such as the power or X-ray beam and the size of radiography rooms puts limitations on the distance between the source of X-rays and the patient. Radiographs performed as portable examinations on a hospital ward or the intensive care unit often require much shorter distances to be used.

As we saw in Chapter 1 specific, characteristic views of particular body structures are a standard part of radiographic practice. When considering some parts of the body such as the knee, elbow, and hand, there will be little alteration in the appearance of the final image by reversing the positions of the X-ray source and the detector. For example, a lateral view of the knee will look near identical if the detector is to the right of the patient and the X-ray beam to the left or the other way around, and likewise these could be reversed from the front and rear positions to perform a frontal view of the knee without making much difference to the image, provided the relative positions remain unaltered (such that the detector plate is always positioned close to the patient). However, for chest radiographs the situation is different. As a result of the heart being positioned at the front of the chest and the spine being located at the back, the projectional nature of radiographs means that there are significant differences in the appearance of a chest radiograph performed with the patient facing towards the X-ray camera (an anterior-posterior or "AP" view) compared to facing away from it (a posterior-anterior or "PA" view). The "PA" projection is considered best practice in most situations, but depending on the age, mobility, and clinical status of a patient an "AP" may need to be used as an alternative, particularly for portable radiographs undertaken on the ward or intensive care unit. The AP view alters the appearance of the ribs, scapulae, and most importantly the heart which is magnified compared to the PA appearances and radiologists need to adjust their interpretation accordingly. Likewise, whilst radiographers will go to great lengths to ensure that patients are positioned straight relative to the axis of the X-ray beam this cannot be achieved in every radiograph and so the radiologist will need to take this positional or projectional distortion into account.

However, the main challenge of interpreting plain radiographs is not in making sense of these (usually modest) positional distortions, but in unravelling the multiple bodily structures superimposed upon one another – the three dimensions of the human frame condensed into a two-dimensional image. We shall see that cross-sectional imaging techniques developed in the late 20th century provide an effective solution to some of these issues, but first, we will consider some related artistic innovations and traditions which either grappled with or help to illustrate these difficulties.

FROM POINTS OF VIEW TO ARROW POINTS

Founding the medical journal *The Lancet* in 1823, politically engaged medical practitioner Thomas Wakley stated "A lancet can be an arched window to let in the light or it can be a sharp surgical instrument to cut out the dross and I intend to use it in both senses".[7]

I have already likened the revelatory nature of medical images to windows and as we continue we shall consider analogies comparing a variety of images to weapons or bladed instruments, utilising Wakley's dual – and definitely not mixed – metaphor. In doing so we shall start with the iconography of the Christian martyr St Sebastian.

The most common and iconic portrayal of St Sebastian is at the moment of his "sagittation" – with numerous arrows piercing his flesh. I have used the term "sagittation" rather than martydom as the saint actually survived this episode of severe trauma, although he was eventually clubbed to death at the orders of the emperor Diocletian. Sebastian's portrayed age and physique evolved over the centuries – in life he was a middle-aged, muscular soldier and this was reflected in early representations, but popular artistic portrayal shifted to a younger, more slender figure over time, also with an increasing proportion of bare flesh exposed.[8,9]

There has also been substantial variation in the number of arrows shown, though without any emerging consistency. Some artists have opted for a parsimonious two or three, half-a-dozen seems the median, whilst some get thoroughly carried away with over 30 in Benozzo Gozzoli's fresco (illustrated in Figure 5.3). A medieval account suggests that by the time the archers were finished the martyr "was as full of arrows as an urchin",[10] urchin here being the archaic English word for hedgehog, rather than a sea urchin,[8] and Gozzoli's depiction certainly bears some resemblance to a hedgehog or porcupine. By contrast, some versions omit arrows altogether portraying a young male figure in a loincloth tied to a tree stump, enough of the iconography provided to allow the figure to be identifiable as Sebastian, and the absence of arrows perhaps heightening tension in anticipation of their arrival.

In those versions in which numerous arrows are deployed it is interesting to note the wide range of directions from which the arrows come. Although thankfully no expert in organising archery-based executions, if I were in charge I would definitely get them all on the same side of the target, rather than firing arrows in the direction of one another. In Figure 5.4, by Michael van Coxie, the arrangement of archers looks particularly hazardous – for the archers in question as much as for St Sebastian.

Compare the painting of St Sebastian by Santa Cruz (shown in Figure 5.5), in which the arrows all appear to have come from the same direction. While the execution squad appear to have been more sensibly positioned on this occasion, as a viewer the arrangement of arrows somehow looks wrong – I just do not feel convinced by the unidirectional configuration of the arrows. Even if the well-being of the archers is compromised, it feels more correct to have arrows originating from numerous different directions. For the full brutality and anguish of the Saint's "first martyrdom" to be conveyed it seems that multiple different arrow trajectories are required.

If we now imagine ourselves in the position of the archers as they take aim from various positions, we appreciate that each of them has a unique perspective on the saint and that the scene could have been rendered from any of these views. Some art historians suggest the popularity of St Sebastian as an iconic template for artists partly relates to the seeming complicity of the viewer in the act of sagittation.[11] By viewing the saint at this moment of extreme vulnerability we are said to participate in the horror of the scene – we may not be piercing his flesh with arrows, but our gaze is equally penetrating.

In more recent times artists have continued to draw upon the iconography of St Sebastian, notably in the cover photograph of the April 1968 issue of *Esquire* magazine featuring Muhammad Ali. Photographer Carl Fischer directly matched Ali's pose and the arrangement of arrows to match the oil painting of St Sebastian by Francesco Botticini (shown in Figure 5.6). Fischer is wise to have chosen a painting with a relatively modest number of arrows depicted. Positioning the arrows and keeping them stationary proved particularly troublesome, requiring a fishing line suspended from the studio ceiling to hold them in the correct positions.[12,13]

FIGURE 5.3   Fresco depicting The Martyrdom of St Sebastian by Benozzo Gozzoli (1465) in the Collegiata of San Gimignano, Italy. Photo by jorisvo. Image credit: Shutterstock

FIGURE 5.4   Martyrdom of St Sebastian. oil painting by Michael van Coxie, St Rumbold's Cathedral, Mechelen, Belgium. Archers are positioned on all sides of St Sebastian, with arrows pointed at the saint, but at risk of hitting the other archers should they miss their target. Photo by Renata Sedmakova. Image credit: Shutterstock

FIGURE 5.5 Oil painting of St Sebastian by Santa Cruz, in which all the arrows embedded within the saint's body appear to have come from the same direction. Credit: Saint Sebastian. Oil painting by Santa Cruz. Wellcome Collection. Attribution 4.0 International (CC BY 4.0).

FIGURE 5.6 St Sebastian by Francesco Botticini. Image Credit: Gwynne Andrews, Rogers, and Harris Brisbane Dick Funds, 1948. The Metropolitan Museum of Art.

Even if the number of arrows is restrained by Gozzoli's standards, the combination of multiple arrows coming from different directions provided a powerful visual metaphor for public attitudes to Ali at the time. He was variously under attack by the American establishment for his race, religion, and refusing the draft to fight in Vietnam. The varied trajectories of the arrow conveys "attacked from all sides" highly effectively, and by surveying the scene from a position amongst the (admittedly non-existent) archery squad the viewer may feel complicit in the onslaught.

PHOTOSCULPTURE

Sebastian's archery squad and sports fans in the stadium remind us that there are innumerable positions from which to observe the action, or come to that any structure being considered. In Chapter 1, we recounted the myth of Dibutades tracing her lover's silhouette from a solitary view, and in the process inventing drawing and painting. Her father is said to have built on this by shaping clay over the outlined profile and in doing so inventing relief sculpture. In the 18th and 19th centuries, there was renewed interest in expanding two-dimensional images into three-dimensional space, and in exploiting multiple viewpoints to do so.

Hand-drawn silhouettes in lateral profile had become a popular means of capturing the likeness of an individual during the 18th century, facilitated by screens such as that illustrated in Figure 5.7. The arrangement is strikingly similar to that shown in depictions of Dibutades (Figure 1.4, Chapter 1). However, in acquiring the trace of the silhouette from the other side of the screen we are faced with the question of whether the resulting image is a truer likeness as rendered from in front or behind the screen (the direction of the sitters gaze being reversed by either choice). Would the resulting image look any different if the sitter had been facing the other way and the resulting outline then been reversed? Even when the choice of where to view the action from appears to be self-evident, niggling questions like these keep cropping up. Either way, the resulting image would look – allowing for individual variation – largely similar to Figure 5.8. In the early days of radiology, radiographs were also referred to as skiagrams (and the first established journal of radiology, later to become the *British Journal of Radiology*, was originally called *Archives of Clinical Skiagraphy*).[14,15] Prior to 1895, *skiagram* had meant "a figure formed by shading in the outline of a shadow",[16] which could certainly be used to describe such silhouette profiles.

Figure 5.9 shows a line drawing of a device called a physiognotype, patented in 1836 by the French inventor Pierre-Louis-Frederic Sauvage, and designed to accurately capture a three-dimensional record of an individual's face. Within an oval metal housing, numerous thin metal rods of equal length were positioned in parallel with one another (similar to the more recent pinscreen executive desk toy), with rods exposed at one end and a taut membrane covering the other side. By gently pressing the exposed rods onto a sitter's face, their likeness was produced in the membrane, from which a wax moulage was produced, in turn used to cast a plaster relief of the face.[17] Although adding an additional step to the pre-existing technique of acquiring a plaster cast of an individual's face, the device made it substantially faster and less messy for the sitter. In the context of antecedents to medical imaging techniques, the use of evenly positioned pins offers another example of a quantised, grid-based image capture technology.

The advent of photography in the second half of the 19th century brought an end to the profile silhouette as a popular phenomenon and meant a short shelf life for the physiognotype. However, building on the endeavour of a repeatable

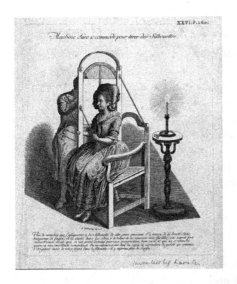

FIGURE 5.7 Eighteenth-century etching showing a seated woman having her silhouette profile drawn by a man using a frame attached to the chair, using candlelight. Credit: A man drawing the silhouette of a seated woman on translucent paper suspended in a frame and lit by a candle. Etching by J.R. Schellenberg, 1783. Wellcome Collection. Public Domain Mark.

FIGURE 5.8 Silhouette portrait of a woman's head and shoulders in profile. Credit: Ada Misner, secretary to Sir Henry Wellcome. Silhouette. Wellcome Collection. Attribution 4.0 International (CC BY 4.0).

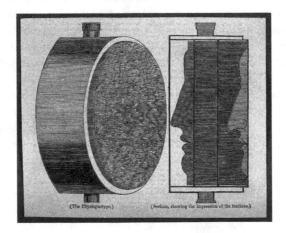

FIGURE 5.9 A line drawing of a device called a physiognotype, used for recording facial profiles, and a cross-section showing the impression of the features. Wood engraving with letterpress, 1837. Credit: The physiognotype for recording facial profiles, and a cross-section showing the impression of the features. Wood engraving with letterpress, 1837. Wellcome Collection. Attribution 4.0 International (CC BY 4.0)

process for capturing a three-dimensional likeness, French sculptor and photographer François Willème developed a technique for producing volumetrically accurate sculptures of the body, which could also be used for accurately reproducing existing sculptures remotely. This was initially called "mechanical sculpture" and then "photosculpture". The technique consisted of taking photographs of a subject using up to 50 cameras positioned in a circle around the individual, not unlike the more reckless archery squad formation encircling St Sebastian. The resultant photographs were then used to outline the profile of the subject onto sheets of wood, metal, or stone using a pantograph. These sheets were cut along the outline producing a negative and positive profile. The sheets of the positive profile were assembled into a three-dimensional array, producing a volumetrically accurate sculpture of a sort in its own right.[18] An example is shown in Figure 5.10. The negative sheets were assembled into a cylindrical structure with a hollow centre, which was used to form a mould from which the final statue was cast. This process, in which – as art historian Robert A. Sobieszek points out – the volumetric whole of a sculpture was considered equal to the sum of its profiles, enjoyed a brief period of success and celebrity in the 1860s but seems to have fizzled out soon after. The technique is, however, thought to have inspired Auguste Rodin who is known to have inspected the silhouette of models from all angles when preparing his sculptures and stated his ambition as being "to capture life by the complete expression of the profiles".[19]

Photosculpture has been recognised as an early precursor of modern 3D printing.[20] It seems the production of photosculptures was largely confined to Willème's studio in Paris, but the technique he pioneered could, nevertheless, allow a subject to be photographed in one location and (after sending the photos or negatives) the sculpture be produced in an entirely different location. Writing this section having just finished a weekend on call in which I was looking at examinations of patients performed in Dundee, Kirkcaldy, and Livingston as well as within Edinburgh all from one reporting station, it is clear that the transferable nature of anatomical representation embodied by photosculpture is now well established in medical imaging.

I would also suggest that in producing a topographically accurate representation of the body by integrating information from numerous separate imaging angles, photosculpture can also be considered a forerunner of sorts to computed tomography. While not revealing the interior, the interrogation of the body from a 360-degree array of imaging detectors is conceptually very similar to CT. Take a look at Figure 5.11. It is a montage of images produced by Nigel Allinson at the University of Lincoln, showing how a single CT image is produced by combining data from multiple X-ray projections. In this case, the images are not those of a live human subject but are a "phantom" consisting of a human skull filled with material to replicate soft tissue. As the number of projections increase from top left to bottom right the initially abstract representations start to become

FIGURE 5.10   Photosculpture – bust of a woman's head composed of numerous wooden cut-outs arranged in a radial configuration. Image courtesy of the George Eastman Museum.

more recognisable. Eventually, the shape of a human skull emerges, much like Holbein's anamorphic skull takes shape as we approach the frame of the painting. In this montage, we have nine images to summarise the process, but the final bottom right image was constructed using 180 different projections in total. To get a feel for the number of individual projections a CT image is produced from, take a look back at Figure 3.12 in Chapter 3 – each linear streak you can see corresponds to an individual X-ray projection. Turning back from that CT image to the photosculpture shown in Figure 5.10, the shared conceptual heritage becomes clear.

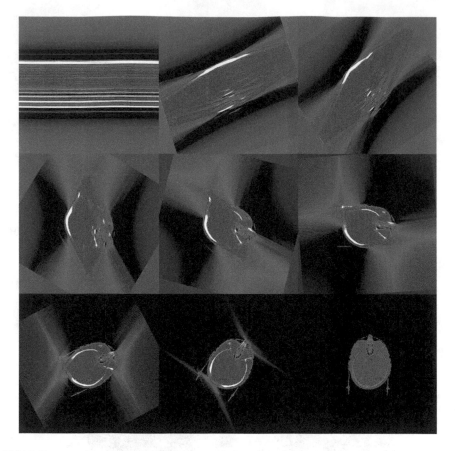

FIGURE 5.11 Montage of images showing how a single CT image is produced by combining data from multiple X-ray projections. In this case, the images are not those of a live human subject, but are a "phantom" consisting of a human skull filled with material to replicate soft tissue. Credit: Steps in producing a proton CT of a human phantom head. Nigel Allinson, University of Lincoln. Attribution 4.0 International (CC BY 4.0).

MULTIPOINT PERSPECTIVE, CUBISM AND THE PARALLAX VIEW

In photosculpture, two-dimensional images were expanded into three-dimensional space, but in plain radiographs the reverse occurs, resulting in superimposition and flattening. Have a look at Figure 5.12 and the abstract nature of squashing three-dimensional structures into a two-dimensional image becomes evident. This is a halftone reproduction of a radiograph, but when I found this image the accompanying label was not immediately available and it took me a moment to recognise that this is a radiograph of a pregnant woman. The lumbar spine of the woman is certainly a familiar appearance to me, located on the left-hand side of the image, being essentially the same appearance of an adult lateral lumbar spine radiograph today. The fetus, in utero, to the right side of the image is harder to make sense of. It is now extremely unusual or unheard of for radiographs to be performed on pregnant women in view of potential harm of the radiation dose to the fetus, so I do not see images of this sort in my clinical practice (ultrasound and MRI now offering superior imaging without ionising radiation). In addition, the foetus is (characteristically) curved up such that various body structures are superimposed over one another, the skull appearing to be detached from the spine on first inspection but most likely relating to a foreshortened appearance of the cervical vertebrae.

When we compare this image to Figure 5.13, *Still Life with a Guitar* by Juan Gris (1913), there is a definite stylistic similarity. Gris called his style Synthetic Cubism, with a characteristically flattened appearance of the objects depicted. By technical necessity, all radiographs produce a flattened silhouette, but the resemblance to the cubist movement is stronger in this case due to the use of shading-like effects in the halftone reproduction of the radiograph. Gris had built on the innovations of cubist pioneers such as Picasso and Braque, who rejected the illusionistic nature of traditional western art, rendered from a solitary viewpoint. Cubist art sought to capture the nature of a structure or subject by displaying numerous different aspects simultaneously in the one work – a similar enterprise to photosculpture, but confined to the two dimensions of a canvas. Picasso is reported to have looked at paintings by Raphael and pointed out that it was not possible to determine the length of an individual's nose due to perspectival distortion.[21] When we look at an example of Picasso's own work (Figure 5.14) we might also struggle to know where to put the calipers. Indeed, the silhouette artists of the 18th century are likely to have come closer to accurately capturing the nasal dimensions of their subjects than any cubist works. However, to paraphrase Paul Klee, art is not in the business of reproducing the visible, but in making things visible.[22] The success of the cubist movement lay not in developing a better alternative to portraying people's noses, but in confronting the impossibility of ever fully capturing the reality of a nose – or any 3D structure – in a two-dimensional image. X-rays, of course, reveal structures previously hidden from view and the revelatory aspects of this emerging technology were not lost on key figures of the cubist movement.

Picasso is also said to have been rather sniffy in relation to how Cezanne portrayed noses,[21] so to side-step nasal controversies let us instead take a look at *Still Life with Apples and a Pot of Primroses* (Figure 5.15), painted by Cezanne in 1890, five years prior to the discovery of X-rays. As with many of his still-life paintings, it looks like a highly unstable arrangement, in which most of the apples look destined to roll off the table, perhaps taking the pot of primroses along with them. This partly relates to the unusual configuration of the tablecloth, which looks as if someone has had a half-hearted stab at the magician's tablecloth removal trick, but is also – more pertinently – related to Cezanne's characteristic use of multifocal perspective.

In many of Cezanne's still life paintings we are confronted with a similar scene; multiple items of fruit plus another larger item such as flowers in a vase sat on a

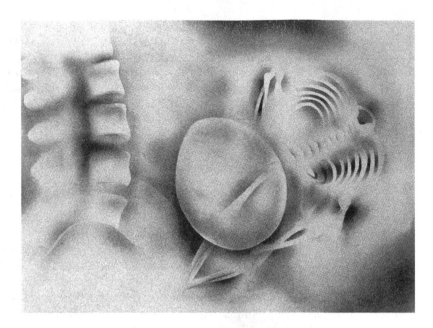

FIGURE 5.12   Halftone reproduction of a radiograph, showing the lumbar spine and abdomen of a pregnant woman, the skeleton of the foetus visible characteristically curled in utero. Credit: Halftone drawing of X-ray. Wellcome Collection. Attribution 4.0 International (CC BY 4.0).

FIGURE 5.13   Still Life with a Guitar by Juan Gris (1913). Everett Collection. Image credit: Shutterstock

FIGURE 5.14   Pablo Picasso – Woman Leaning 1, 1939. Portrait of a woman leaning her face on her right hand, but in a cubist style such that the relative positions of her eyes appear anatomically implausible, with the left eye hovering over the anticipated area of her forehead, and the nose appearing very distorted, with each nostril heading in opposite directions. Image credit: Shutterstock

FIGURE 5.15   Still Life with Apples and a Pot of Primroses, painted by Cezanne in 1890. Image credit: Everett Collection /Shutterstock

table, often with a very badly arranged tablecloth partly covering the table, but with a similar multifocal perspective. It is as if we are viewing each scene from the point of view of someone sitting in front of the table and standing in front of the table simultaneously. Many of the apples seem to reveal both the top surface, with the pit for the stalk clearly depicted, and the side. These fruits, when scrutinised individually, somehow look the wrong shape, perhaps stretched a little top to bottom. Yet within the scene as a whole they blend into this visual milieux, not looking out of place and coherent within the image. Likewise, the table has a paradoxical appearance – the flat surface of the table appears elevated towards the viewer, as if the rear table legs were taller than the front ones and the table surface was slanted. In a more extreme form the distortion of space from the conventional rendering of linear perspective starts to resemble the scene in the high-concept thriller *Inception* (dir. Christopher Nolan, 2010) in which the streets of Paris are folded back upon themselves, also echoed in the less high-concept *Detective Pikachu* (dir. Rob Letterman, 2019) which also features a scene in which the landscape curves such that the hillside scenery is both under the characters' feet and over their heads.

Cezanne's multifocal synthesis utilises a roving, dynamic viewpoint, conceptually very similar to techniques developed by early radiographic pioneers to help locate embedded metal foreign bodies within the body. Exploiting the parallax principle, multiple oblique views were performed to determine the depth and position of bullets or shrapnel fragments beneath the skin surface. The technique is well described in numerous 19th-century textbooks of radiography and was utilised extensively during the First World War.[23]

A roving viewpoint is also the fundamental principle of focal plane tomography, a radiographic technique developed in various iterations during the 1920s and 30s by radiographic pioneers, including Andre Bocage, Alessandro Vallebona, and William Watson.[24] The X-ray source and film were moved in synchrony in opposite directions to each other, resulting in structures located in a centrally located focal plane remaining sharply delineated, whilst structures located outside this plane become blurred.[25] The technique achieved some degree of success in relation to the problem of superimposition, but mainly in the realm of bony or densely calcified structures, as soft tissues within the focal plane tend to merge with the blurred soft tissue elsewhere. One tomographic technique still used in clinical practice today is the orthopantomogram (OPG), illustrated in Figure 5.16. In this example, the motion of the X-ray source and detector results in the projectional artefact of seeing four earrings instead of two and two nasogastric tubes instead of one. Within my own department the OPG is the only radiographic technique that uses focal plane tomography, and as such is something of a rarity, although digital tomosynthesis may lead to a revival of such examinations in other settings.[26] Nevertheless, most plain radiographic tomographic techniques have been superseded by the development of CT and MRI in the final decades of the 20th century, techniques that achieved a truly unobscured window into the body's interior.

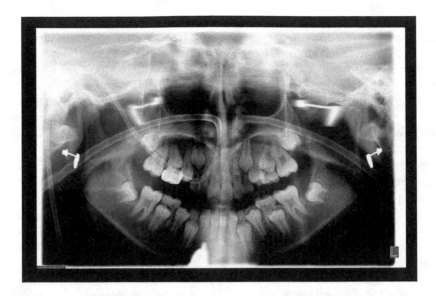

FIGURE 5.16   Orthopantomogram (OPG). A tomographic technique allowing the teeth and mandible to be visualised without superimposition, but resulting -in this example - in the artefactual appearance of four earrings being visible instead of two and two nasogastric tubes instead of one. Image provided by the author.

CROSS-SECTIONAL IMAGING AND MULTIPLANAR REFORMATTING

A full account of the origin of CT and MRI scanning technology is well beyond the remit of this book, but it is notable that two of cross-sectional imaging's undisputed heroes both left school with no qualifications. Sir Godfrey Hounsfield, who built the first functioning CT scanner in 1972 and was awarded the Nobel Prize for medicine based on his efforts in 1979, left school at 16 with a report card suggesting his poor schoolwork was the result of intellectual retardation.[27] Sir Peter Mansfield failed the 11-plus examination, left school at 15 with no formal qualifications and was instructed by his career adviser to consider something less ambitious than a career in science. In 2003 he was awarded the Nobel Prize for medicine in recognition of his contribution to the development of MRI, including developing the first scanner to acquire images of a living human subject.[28]

While Paul McCartney passed the 11-plus and subsequently got A-levels in English and Art, the rest of *The Beatles* followed in Hounsfield and Mansfield's footsteps, leaving school without any formal qualifications. The urban myth goes that EMI (the company in whose laboratories Hounsfield was working in the late 1960s and 70s) had more money than they knew what to do with in those years and funded the development of the CT scanner as part of a speculative "blue-sky" research programme off the back of *The Beatles'* record sales. This version of events has been dispelled in *Godfrey Hounsfield: Intuitive Genius of CT*, which documents the stretched budget of the project. Hounsfield was only given a £5,000 budget to build a prototype CT scanner, rather than the £20,000 he had asked for,[29] and there was a reduction in EMI research staff of 25% in the years the first scanner was built.[30] On this basis, however, I'm going to keep *The Beatles* in the CT story after all as it is possible the company may have folded altogether had it not been for the record sales! Either way, we should certainly value the potential of all individuals regardless of their academic achievements at school.

While generated using very different mechanisms, both CT and MRI produce planar images of the body allowing structures to be seen topographically rather than superimposed over one another as is the case with plain radiographs. These modalities are often referred to as cross-sectional imaging due to the resemblance of slicing through a structure and then looking at the exposed ends. Figure 5.17 shows an axial CT image of the chest, alongside Figure 5.18, an illustration from a 19th Century anatomy textbook at approximately the same location. While ultrasound effectively produces images in cross-section, the quality of the image typically deteriorates with depth from the skin surface, and for curvilinear probes the image appears stretched as the beam diverges, not unlike Holbein's skull. In addition – as mentioned earlier – structures can be obscured by bowel gas or bony structures, so ultrasound is therefore not usually included as a cross-sectional modality. In employing a grid system akin to a map or woven cloth, cross-sectional imaging solves the problem of superimposition, and in doing so this imaging technology has saved countless lives over the past half-century. However, it is worth noting that for any single image the coverage is always incomplete, and the issue of viewpoint is only partially solved. Like the archers taking up their positions around St Sebastian, decisions still need to be made concerning where the action will be viewed from. Consequently, images are constructed presupposing a very particular point of view. Cross-sectional images can be topographically accurate to a submillimetre scale and, as such, are not illusionistic in the sense attributed to perspectival works. They nevertheless do remain a highly abstracted representation of the body.

Historically, CT scans were known as CAT scans standing for Computerised Axial Tomography, but over time this evolved into computer-assisted tomography, before the current shorter version of computerised tomography became favoured. So why did axial get dropped? For the first 30 years or so of CT scans the images were predominantly acquired in axial section, and exclusively so for most of that period. This was for practical reasons – scanning through the smallest diameter of the body enables a smaller gantry size and enables effective localisation of the scan to the area of interest. The first CT scanner was designed specifically for scanning heads, facilitating an even smaller gantry diameter, as well as offering visualisation of that body region largely hidden from X-ray-based imaging until that point – the interior of the skull.

However, over the past 20 years, multidetector CT scanners have altered the way in which CT datasets are acquired, facilitating volumetric scans in which the information can be reconstructed in any anatomical plane. With the exception of some specific applications such as CT-guided biopsies, the large majority of modern CT

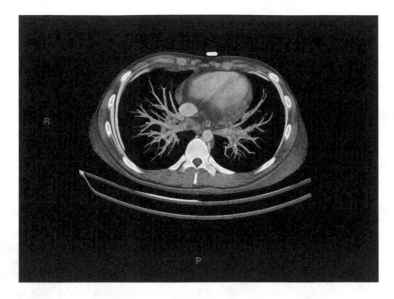

FIGURE 5.17   An axial CT image of the chest. Image provided by the author.

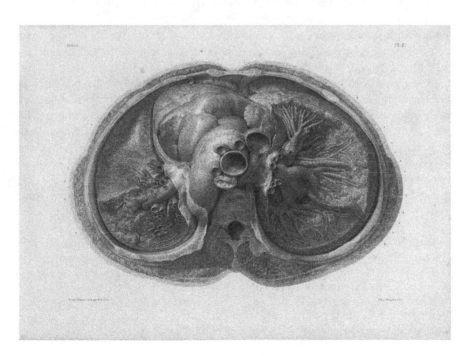

FIGURE 5.18   An illustration from a 19th-century anatomy textbook, showing the chest in axial section. Bourgery, J.M. 1797–1849. Credit: *Traité complet de l'anatomie de l'homme comprenant la médecine opératoire*. Wellcome Collection. Attribution 4.0 International (CC BY 4.0).

examinations are conducted in a way that no longer confines image interpretation to the axial plane.

The introduction of MRI in the late 1970s and early 1980s had already enabled anatomical planes other than axial to be imaged, most notably sagittal imaging of the brain, crucial for the evaluation of midline structures such as the pituitary gland. Despite this, for many years the plane of imaging was constricted to those imaging planes preselected for the examination, such that once the scan was concluded the radiologist was stuck with images acquired in those planes only. Subsequent advances in scanning technology have allowed the MPR (multiplanar reformatting) mode of both CT and MRI to become a standard tool of the radiologist, allowing the scan to be displayed in any anatomical plane, long after the patient has left the department.

In this mode, the screen is typically divided between one large pane in which the desired, tailored plane of imaging is displayed, and three smaller navigational panes in which the scan is shown in axial, sagittal, and coronal planes (as shown in Figure 5.19). To contemporary readers this arrangement is quite reminiscent of video-conference calls, in which meeting participants are demarcated within the rectangular windows. It also reminds me of the screen layout in the film *Timecode* (dir. Mike Figgis, 2000) in which the cinema screen was subdivided into four rectangles from which the "real-time" story unfolds from four different viewpoints (each an impressively staged continuous tracking shot, which eventually converge upon each other). Just as different participants in the video call may provide complementary contributions, or examining a narrative from different characters' perspective may enrich a drama, so the scrutiny of structures of clinical interest from multiple planes of inspection can enhance the level of information a radiologist can provide.

However, the shift from having one solitary anatomical plane to look at to every conceivable plane poses some challenges. Video calls can become unmanageable with too many participants and despite the technical brilliance of *Timecode* the simultaneous four viewpoint narrative device did not catch on in a big way. Likewise, it is possible to have "too much of a good thing" in relation to the multiplanar interrogation that modern cross-sectional imaging allows.

In *The Human Touch*, Michael Frayn speculates on what divine vision might entail:

> [God] saw it all in one go, continuously and eternally. And since God was everywhere he saw it not in perspective, not from some particular viewpoint, but from every possible viewpoint. From all sides of a cube simultaneously, for example. From an angle of ninety degrees to each of those sides - from an angle of one degree, eighty-nine degrees, seventeen degrees. From a millimetre off and a mile off. From every point inside the cube looking out. He could see up your trouser-leg and down your trouser-leg. See your vest through your shirt and your chest through your vest.[31]

Radiologists may be prone to delusions of grandeur from time to time. Revealingly, one of my colleague's favourite jokes is; "How do you kill a radiologist? Get them to jump from their perceived knowledge base to their actual knowledge base". However, most would acknowledge that the capabilities of modern imaging techniques fall

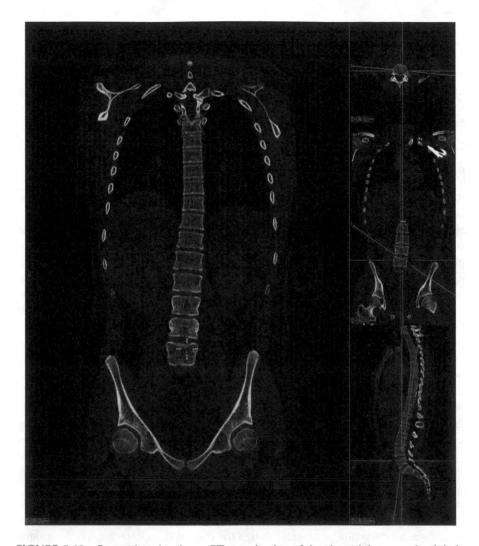

FIGURE 5.19 Screenshot showing a CT examination of the chest abdomen and pelvis in the multiplanar reformatting tool display. The three smaller panels on the right-hand side show the images in axial, coronal, and sagittal planes. The larger pane shows the reformatted image in the plane selected by the radiologist – in this case coronal plane but "straightened up" compared to the original examination (note the more symmetrical appearance of the hips compared to the thumbnail coronal image). Image provided by the author.

some way short of god-like powers. In Frayn's description God is visualising our individual not only from every possible angle, but doing it all simultaneously (as well as visualising everything else, everywhere). The rate-limiting step in the X-ray department is that whoever is viewing the images is only human and can only process one thing at a time.

Nevertheless, with that "one-thing-at-a-time" caveat established, this description is not so very far away from what modern imaging techniques allow. The MPR mode enables not only the three conventional anatomical planes to be displayed from a single volumetric acquisition but everything in between (whether 1 degree, 17 degrees, or 89 degrees). Indeed, through the use of animated 3D reconstructions, the option to go up one trouser leg and down the other is potentially achievable – if unlikely to be desirable (or indeed ethical).

POV

Rather than exploring people's trouser legs, more practical use of this technology is achieved by going on a virtual 3D tour of a patient's airways (virtual bronchoscopy) or large bowel (virtual colonoscopy) in which the anatomy is displayed in the style of a first-person shooter videogame, though perhaps more akin to a potholing or caving expedition. The similarity to videogames is not entirely coincidental, with a big overlap in the software/visual processing algorithms used in the gaming industry and medical imaging. This characteristic "fly-through" point of view produces a similar effect to the viewer having been shrunk down to a very small size, exploring the body in a submarine or spaceship style vehicle as portrayed in the movies *Fantastic Voyage* (dir. Richard Fleischer, 1966), or *Innerspace* (dir. Joe Dante, 1987). Virtual colonoscopy (illustrated in Figures 5.20 and 5.21) therefore owes a debt to both cinema and to videogames. While movies have provided the majority of visual pop culture reference points for the 20th century, videogames are increasingly making inroads in this direction, such that several film critics made the comparison of continuous tracking shots following British soldiers around the trenches in World War I drama *1917* (dir. Sam Mendes, 2019) to being in a first person shooter.[32,33,34] It is increasingly difficult to know who is influencing whom in this cyclical process.

Within cinema, at any rate, the term "point of view" (or POV), specifically refers to camera shots acquired from an angle and height that mimic the viewpoint of a particular character, and if achieved successfully the audience will understand that we are seeing the action as though through the eyes of that character. The technique has also been heavily utilised in television, particularly in the long running UK sitcom *Peep Show*. Film-maker and writer Mark Cousins describes POV as "cinema's party-piece – it's everyday astonishment".[35] The astonishment factor can be magnified when instead of a character we see the action from the POV of a speeding projectile such as a bullet (*Sniper*, dir. Luis Llosa, 1993; *Lord of War*, dir. Andrew Niccol 2005) or – perhaps you are picking up on a recurring motif – an arrow (*Robin Hood: Prince of Thieves*, dir. Kevin Reynolds, 1991).

A familiar variation on the POV shot is where we see the viewpoint of a character looking through some sort of long-range visual device such as a telescope, periscope, or gunsight. For binoculars we are used to seeing a figure-of-eight shaped black border around the edge of the image, a now well-established device even if not particularly akin to the actual experience of peering through binoculars. Film-makers have also actively employed the black margins at the edges of

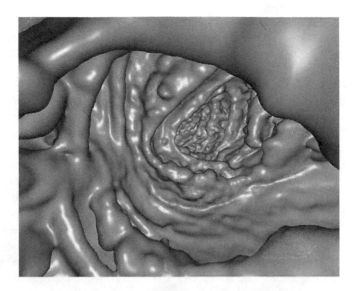

FIGURE 5.20   Virtual colonoscopy. Note the use of directional light to produce reflections and shadow effects. Image courtesy of Dr A Green.

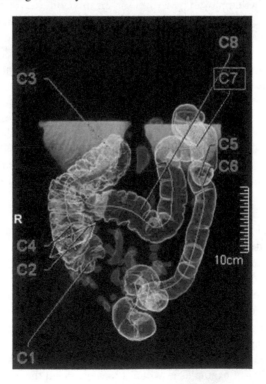

FIGURE 5.21   Virtual colonoscopy roadmap. The course along which the "fly-through" view of the colon follows can be substantially manipulated by the radiologist. Image courtesy of Dr A Green

the image for other purposes.[36] A familiar example is the start of a film mimicking the narrow aspect ratio of a traditional television screen (1.33:1) before opening up to widescreen (2.35:1), employed by celebrated directors including Oliver Stone (*JFK*, 1991), Brian De Palma (*Snake Eyes*, 1998) and Steven Spielberg (*Catch Me If You Can,* 2002). CGI movies including *The Incredibles* (dir. Brad Bird, 2004) and Up (dir. Pete Dochter, 2009) also employ modification of the aspect ratio to similar effect. In *Da 5 Bloods* (dir. Spike Lee, 2020) the aspect ratio changes several times throughout the movie, a transition to the narrower, near-square proportions indicating a flashback scene. Films which had a 3D theatrical release sometimes make use of the horizontal black bands that maintain the widescreen aspect ratio when converted to narrower television screens. Fish appear to jump out of the screen by traversing the lower horizontal band, unconfined by the usual rectangular confines in *Life of Pi* (dir. Ang Lee, 2012) and the *Minions* (dir. Kyle Balda, Pierre Coffin, 2015) also refuse to be restricted by this convention when viewed on DVD.

Outside of the immediate clinical environment, this ability to control the horizontal and vertical margins of the image does sometimes take on a more artistic or at least theatrical dimension. During my training it was impressed upon me that the radiologist has a responsibility to look at everything included in the image, and an established means of determining if trainees are adhering to this is to include "edge-of-film" findings in examination cases. Such findings (perhaps an ingested foreign body in the upper abdomen only just visible at the bottom of a chest radiograph, or a fractured bone right at the edge of the image) do occur in "real-life", providing a resource of teaching cases, but quite often a finding of this nature is not quite close enough to the edge to make it a challenging spot. Within my own teaching collection I have a number of examples in which I have cropped the image to deliberately conceal portions of the relevant finding or at least bring it closer to the edge of the displayed image (or the outer limits if you prefer).

In each of these situations, we are reminded of the window-like constraints through which we observe the action. As a frustrated wannabe film-maker I certainly think of the collimation, choice of angle and duration of screening I utilise when performing a dynamic fluoroscopy examination as being akin to the shot selection and editing employed by a director. In this context, focusing the "action" within the centre of the frame is less about an artistic imperative and very much about the necessity to reduce radiation exposure (the smaller the field of view, the smaller the radiation dose). The selectivity of the windows through which events unfold in fluoroscopic examinations such as barium swallows and follow-throughs does, all the same, have a salient similarity to the time-travelling and viewpoint changing aspects of cinema – a property also shared by ultrasound examinations. In the interests of full disclosure, collimation of fluoroscopic images can be modified using a variety of apertures including circular and oblique parallelograms. Despite this, rectangular columns are most typically used, and in controlling these up and down and side-to-side margins I am reminded of the opening words to the TV series *The Outer Limits* in which the narrator promises to control both the horizontal and the vertical.[37]

Outside of the immediate clinical environment, this ability to control the horizontal and vertical margins of the image does sometimes take on a more artistic or at least theatrical dimension. During my training it was impressed upon me that the radiologist has a responsibility to look at everything included in the image, and an established means of determining if trainees are adhering to this is to include "edge-of-film" findings in examination cases. Such findings (perhaps an ingested foreign body in the upper abdomen only just visible at the bottom of a chest radiograph, or a fractured bone right at the edge of the image) do occur in "real-life", providing a resource of teaching cases, but quite often a finding of this nature is not quite close enough to the edge to make it a challenging spot. Within my own teaching collection I have a number of examples in which I have cropped the image to deliberately conceal portions of the relevant finding or at least bring it closer to the edge of the displayed image (or the outer limits if you prefer).

IF LOOKS COULD KILL …

In her book *On Photography*, Susan Sontag suggests:

> There is something predatory in the act of taking a picture. To photograph people is to violate them, by seeing them as they never see themselves, but having knowledge of them they can never have; it turns people into objects that can be symbolically possessed. Just as the camera is a sublimation of the gun, to photograph someone is a sublimated murder - a soft murder, appropriate to a sad, frightened time.[38]

While Sontag is by no means alone in describing what we might call weaponised image acquisition – "camera shots" and "photo shoots" are revealing long established phrases in this context – she does put it in very striking terms. If a photograph is a sublimated murder, I shudder to think what a cross-sectional imaging examination such as CT or MRI would equate to. We have already referred to CT images by the most commonly used term – "slices" – but these are also known as "cuts". The toolbar I use on the computer when producing 3D reconstructions has a cartoon-like icon of a pair of scissors for removing unwanted parts of the scan. While this might include external structures like ECG leads, most of the time it is parts of the patient's body that the scissors are cutting away. CT and MRI examinations undertaken as post-mortem examinations are often called "virtual autopsy", implying the imaging has taken on the role of a substitute scalpel.

In clinical radiology the intention is to assist the patient, with safety and avoidance of harm central to all activity undertaken in an X-ray department. Likewise, a virtual autopsy may help offer some closure for the family of the deceased, provide information to guide genetic counselling or criminal investigation, and may indeed obviate the need for physical incision of the flesh. These metaphorical allusions to cutting, slicing, and even to murder intuitively feel misplaced.

However, historically the association of dissection as a punishment for criminals, alongside the religious belief in the resurrection of the physical body after death, meant anatomical inspection of the body has long been perceived as an aberration. Sixteen individuals were killed by Burke and Hare in 1828 to provide cadavers for dissection in the anatomy school of Dr Robert Knox in Edinburgh – the means utilised to look inside the body taken to a horrific extreme, murder of a literal, not figurative nature. In *Screening the Body*, Lisa Cartwright points out that the pioneering work of German radiologist Robert Janker conducted in the 1930s includes a fluoroscopy examination demonstrating the appearance of a man's larynx as he speaks. The synchronised audio recording reveals the man is reading a nationalist speech in support of the Third Reich.[39] Anatomical studies would later be enabled on an industrial scale by mass murder under the Nazi regime in the 1940s.[40]

Imaging techniques may not incise the body, and should always be conducted with the patient's consent, but the interpretation of such images draws upon a troubled heritage of anatomical visual representations.[41] While murder victims probably constitute a very small fraction of bodies dissected for anatomical purposes through history, I suspect that consent was not forthcoming for the large majority.

Modern radiology examinations may be safe, but the act of looking may not always be benign. The information provided by antenatal imaging may, for example, be pivotal in determining the continuation or otherwise of a pregnancy.

Have a look at Figures 5.22–5.25. Figure 5.22 is known as "The Wound Man" a characteristic type of surgical illustration seen from the medieval era onwards (this example dating from 1530). Figure 5.23 is from the same text, showing body sites suitable for blood-letting. The labelling lines resemble both the weapons embedded within the flesh of the Wound Man, and the arrows piercing the flesh of St Sebastian, a contemporaneous portrayal of whom is shown in Figure 5.24. There is also a definite resemblance to depictions of St Sebastian, in the *écorché* figure shown in Figure 5.25 – arrows replaced by the vector-like label lines, the text of each label not dissimilar to the fletchings of the arrows. Anatomical artists of this era were well versed with these established religious motifs, so I would suggest the resemblance is not entirely coincidental. In flaying the flesh, exposing the internal structures of the body and labelling them in this distinctive fashion this figure follows in the visual tradition of anatomical martyrs. An *écorché* figure holding aloft its own flayed skin is an anatomical motif in its own right, drawing upon the iconography of another

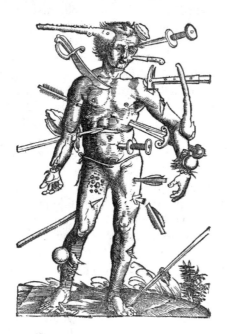

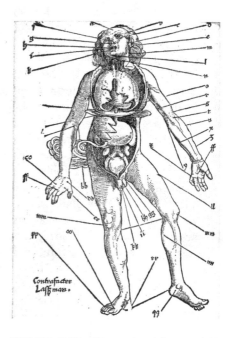

FIGURE 5.22 "The Wound Man" a characteristic type of surgical illustration seen from the medieval era onwards. Credit: Feldtbuch der Wundartzney, newlich getruckt und gebessert/[Hans von Gersdorff]. Wellcome Collection. Attribution 4.0 International (CC BY 4.0).

FIGURE 5.23 Illustration of a partially dissected figure, from the same text as Figure 5.23, showing body sites suitable for blood-letting. Credit: Anatomical blood-letting figure, 16th Century. Wellcome Collection. Attribution 4.0 International (CC BY 4.0).

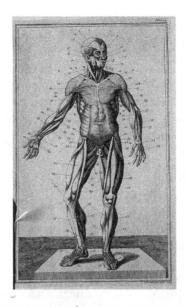

FIGURE 5.24   Woodcut of St Sebastian, flesh pierced by numerous arrows and with archers depicted on each side. Credit: Martyrdom of Saint Sebastian. Reproduction of coloured woodcut. Wellcome Collection. Attribution 4.0 International (CC BY 4.0).

FIGURE 5.25   *Écorché* figure with labelling lines. Credit: The muscles of the human body, seen from the front, after Eustachius. Etching by G. Bickham, 1743, after himself, after an engraving, c. 1552. Wellcome Collection. Public Domain Mark.

Christian martyr, Saint Bartholomew, which we will return to in Chapter 8. In this context, the visual symmetry between embedded weapons and arrows and the labelling lines of anatomical figures is significant. The act of labelling, as we shall revisit in Chapter 6, is a transformative one which may be empowering or pejorative. To label a structure in an anatomical representation it must be visible in the diagram which presupposes a particular point of view, in the same way that for an arrow to pierce the flesh of St Sebastian, the archer requires a line of sight for the arrow to follow. Whole-body figures such as these are largely absent from modern anatomy and radiology textbooks, with images typically restricted to localised body parts or regions, but arrow symbols are still a consistent feature to highlight particular structures, maintaining this tradition.

In *Anatomy Acts*, a series of insightful essays surveying anatomy in Scotland over 500 years, Jonathan Sawday suggests that

> Planting their own names on the body's interior, in much the same way that the explorers and discoverers claimed whole continents in the name of their respective sovereigns, or commemorated their friends and patrons by submitting geography to European systems of nomenclature, the anatomists were participating in that great enlightenment project by which the world, and all that it contained was systematized and regulated.[42]

In this analysis the arrow-like labelling lines also start to resemble flags planted into the territory of the body, indicating ownership and colonisation as much as insight or understanding. Sawday's use of the word "planting" is also salient in this context. The Latin word *condo* can be translated as plant but can be used in very different situations; planting (or building/founding) a new settlement such European explorers might, or planting (or plunging) a sword into an enemy's body. Virgil uses both meanings in describing the deeds of Aeneas in the *Aeneid*.[43] The word "pinpoint" when used as a verb also straddles cartography and anatomy, reminiscent of both the marker icon on online maps and the indicators of the structure of interest in anatomy spot tests at medical school.

Sadly, piercing of the flesh is neither confined to figurative mechanisms, or to distant historical events. In modern life, CT scans are often required to evaluate the injuries of stabbing victims. While such cases are rare in paediatric practice, I recently attended a meeting in which colleagues from a larger urban centre presented a series of cases of teenage victims, displaying images of the different types of knife alongside the CT images of their injuries. Knowledge of the length of the blade and the contour of the knife edge (such as serrations) is important to help ensure the images are interrogated appropriately for deep trauma. The range of different knives and blades that were displayed was disturbingly similar to those seen in Figure 5.22. In cases of such penetrating injuries, the facility to modify the imaging plane to match that of the trajectory of the blade is very useful to determine the exact depth of the laceration and to identify any vascular or visceral injury along its path.

The ability to fine-tune the plane of imaging has a number of other applications. "Axial" MRI images of the spine are usually acquired a few degrees off the transverse plane to match the curvature of the spine, and the axial planes in which MRI and CT examinations of the brain are most typically acquired also vary by a few degrees (the CT plane being skewed by a few degrees such that the scan can avoid exposure of the radiosensitive lenses in the eyes). The MPR function allows the diverging anatomical planes to be reconciled if a patient has both MRI and CT studies that require comparison. Virtual colonoscopy and bronchoscopy pay no heed to the standard anatomical planes (although a lesion identified on these POV type images will usually be correlated with the planar images to make sense of where it is located). We also saw in the previous chapter that the brain can be rolled out flat, and similar manipulation of crosssectional imaging data can be used to artificially straighten scoliotic spines to assist anatomical navigation. 3D rendered reconstructions can be rotated through 360 degrees, enabling the "sum of all profiles" type visualisation promoted by photosculpture.

However, when scrutinising cross-sectional studies, radiologists tend to spend most of their time confined to the conventional planes and of those axial remains the most commonly used. Radiologists have a tendency to gravitate back to axial images like a multilinguist reverting to their first language.

INFINITE POSSIBILITIES – FINITE RESOURCES

Notwithstanding the caveat that I insisted on for human radiologists – we can only look at one thing at a time – the end product of looking at a cross-sectional study (the radiology report) must nevertheless integrate inspection of multiple bodily structures from an innumerable number of angles, quite often using several different imaging modalities and perhaps comparing to a number of previous examinations. All told this may require the analysis of several hundred, if not thousands of individual, data-laden images. Preparation of a multidisciplinary team meeting (MDTM) to discuss a number of complex patients typically involves reviewing tens of thousands of images (How else can you kill a radiologist? Get them to jump off their pile of MDTM images!). The act of interpreting just a single image (let's say a chest radiograph) without reference to any previous imaging draws upon an enormous cache of visual references, whether other CXRs, anatomical dissection and diagrams from medical school, other imaging modalities that help make sense of the anatomy and so forth. It requires the combined assessment of different body structures (heart, lungs, bones, soft tissues, etc.) to make a diagnosis. Even in the evaluation of a single bone radiograph, we have seen that it is usual practice to perform two orthogonal views such that the opinion offered in the radiology report is a synthesis of both images.

Although at the time of writing sports stadia are largely empty in the wake of the pandemic, top-level events continue to be covered by a large number of TV cameras. In recent years the introduction of video-assisted refereeing in Premier League football has recast the role of such cameras from passive observer to active participant in the events of the match. Controversy has ensued both in relation to individual decisions and the process as a whole, with many fans and commentators suggesting the numerous stoppages interfere with the flow and spontaneity of the game. Picasso's interest in nasal morphology is of relevance here – UEFA President Aleksander Ceferin has suggested that having a long nose might put a player offside based on the close margins used in the VAR era.[44] It turns out that looking at a possible handball from numerous different camera angles can, on occasion, make it harder to come off the fence one way or another. My 12-year-old son had some insight when he suggested that "VAR is ultimately a bit pointless as somebody still has to make a decision in the end".

Just as somebody must make the call as to whether to award a penalty, or rule out a goal on the basis of a borderline offside, so the radiologist has to make a call on the basis of imaging. Sometimes the decision of one disease over another can be postponed by repeating the imaging or doing another type of imaging prior to making that call, but sometimes the hardest decision to make is when to stop imaging – deciding not to suggest another imaging test. Even in the context of one cross-sectional imaging examination, with limitless options for planar and 3D reconstructions, the temptation to keep adjusting the image post-processing functions can be strong. The practical consideration of getting through the reporting pile usually prevents too much messing around but in order to "streamline" the viewing process to ensure a thorough evaluation of the images devoid of unnecessary faffing an intuitive approach based on experience is usually required.

SUMMARY

The predominant visual motifs of this chapter have been windows and arrows, so let's see if I can tie it all up using arrow slits as the final analogy. These are the window-like narrow vertical apertures found in medieval era castles through which an archer could launch arrows, whilst largely protected from enemy fire. There is clearly a compromise to be struck in the dimensions of the slit – too wide and the archer is exposed, too narrow and the scope for possible targets is limited. Likewise, in the depictions of St Sebastian artists need to make a choice of how many arrows to include. In medical imaging the information provided by a solitary perspectival or projectional image may be too limited to provide adequate assessment, but at the other end of the scale it is not a good use of resources for radiologists to pore over cross-sectional imaging studies from every conceivable plane. Indeed, sometimes just the one projection *is* sufficient to answer the pertinent clinical question and the role of the radiologist may be to articulate that no further imaging is required, much as a good referee will not resort to video analysis when sure of their verdict. Judgement is therefore required to ensure that studies are examined in sufficient detail to extract all the clinically relevant information but knowing when to move on to the next case. Like Thomas Wakley using his *Lancet* to cut out the dross, the radiologist must cut out not only unhelpful content from radiology reports but also unnecessary imaging.

The conceptual overlap between some of this chapter's themes – penetrating injuries, planar trajectories, and transparency of the flesh – is manifest in a beautiful glass sculpture of St Sebastian by Clifford Rainey, displayed at the National Museum of Scotland, Edinburgh, and illustrated in Figure 5.26. Instead of arrows, planar sections intersect the body at three oblique angles through the torso. These fail to correspond to conventional anatomical planes, but closely resemble the tailored imaging planes made possible by multiplanar reconstruction. Unlike the axial plane, the other two conventional anatomical planes (coronal and sagittal) are named after sutures in the skull located at the junctions of different skull bones. The sagittal suture is said to resemble an arrow (*sagitta* being the Latin word for arrow), and St Sebastian's "sagittation" is also derived from similar origins. This sculpture breaks with much of the conventional iconography of St Sebastian to achieve a fresh take on a familiar subject. Nevertheless, the figure is recognisably following in that well established visual tradition, and looking back at the various illustrations of St Sebastian featured in this chapter together with the martyr-like anatomical figures, the visual parallels start to stack up. It is our collective tendency to rely on what is familiar that we will tackle in the next chapter.

FIGURE 5.26   St Sebastian by Clifford Rainey, displayed at the National Museum of Scotland, Edinburgh. (photograph by the author, with permission from the artist)

REFERENCES

1. Macmillan Dictionary Blog, https://www.macmillandictionaryblog.com/perspective, last accessed 18/07/21.
2. B. Smith, T. Grid, from D. Montello (ed), *Spatial Information Theory. Foundations of Geographic Information Science* (Lecture Notes in Computer Science 2205), Springer, Berlin; New York, 2001, 14–27.
3. J. Brotton, *A History of the World in Twelve Maps*, Allen Lane, London, 2012, p 394.
4. J. Berger, *Ways of Seeing*, Penguin Books, London, 1972, p 16.
5. R. Shiff, Bridget Riley: The edge of animation, in *About Bridget Riley: Selected Writings 1999 -2016*, Ridinghouse, London, 2017, pp 230–231.
6. A. M. Lippit, "Archetexts": Lascaux, Eros, and the Anamorphic subject, *Discourse*, 24(2), Nature Art and Urban Spaces, Spring 2002, 18–29 (12 pages) Published By: Wayne State University Press.
7. J. Pockley, Innovation and reflection, *The Lancet* 175, 352 (October Supplement 2), 1998. SII2.
8. Wikipedia, Saint Sebastian, https://en.wikipedia.org/wiki/Saint_Sebastian#:~:text=Sebastian%20had%20prudently%20concealed%20his,would%20shoot%20arrows%20at%20him.&text=Miraculously%2C%20the%20arrows%20did%20not%20kill%20him, last accessed 18/07/21
9. C. Darwent, Arrows of desire: How did St Sebastian become an enduring, homo-erotic icon? *The Independent*, 10 February 2008.
10. Medieval sourcebook: The golden legend (Aurea Legenda), Compiled by Jacobus de Voragine, 1275, Translated by William Caxton, 1483, Fordham University, https://sourcebooks.web.fordham.edu/basis/goldenlegend/GoldenLegend-Volume2.asp#Sebastian, last accessed 18/07/21.
11. M. Wood, *Black Milk: Imagining Slavery in the Visual Cultures of Brazil and America*, Oxford University Press, Oxford, 2013, pp 156–159.
12. Jill Hudson, Muhammad Ali's 1968 'Esquire' cover is one of the greatest of all time, *The Undefeated*, 31/05/17, https://theundefeated.com/features/muhammad-ali-1968-esquire-cover/, last accessed 18/07/21.
13. Alexxa Gotthardt, The photograph that made a Martyr out of Muhammad Ali, *Artsy,* https://www.artsy.net/article/artsy-editorial-photograph-made-martyr-muhammad-ali, last accessed 18/07/21.
14. Daniel J. Bell, Archives of clinical skiagraphy, *Radiopaedia*, https://radiopaedia.org/articles/archives-of-clinical-skiagraphy?lang=gb, last accessed 18/07/21.
15. Daniel J. Bell, Skiagraphy, *Radiopaedia*, https://radiopaedia.org/articles/skiagraphy-terminology?lang=gb, last accessed 18/07/21.
16. Merriam Webster, Online dictionary, https://www.merriam-webster.com/dictionary/skiagram#h1, last accessed 18/07/21.
17. R.A. Sobieszek, Sculpture as the sum of its profiles: François Willème and Photosculpture in France, 1859–1868, *The Art Bulletin*, 62(4), Dec., 1980, 617–630, p 624.
18. Ibid p 627
19. Ibid p 630
20. Jeremy M. Norman, François Willème invents photosculpture: Early 3D imaging, *History of Information*, https://www.historyofinformation.com/detail.php?entryid=4335 Last accessed 18/07/21
21. E. H. Gombrich, *Topics of Our Time: Twentieth-Century Issues in Learning and in Art*, Phaidon, London, 1991, p 137.
22. A. Montfort, Making visible, in *About Bridget Riley: Selected Writings 1999 -2016*, Ridinghouse, London, 2017, p 359.

23. A. M. K. Thomas and A. K. Banerjee, *The History of Radiology*, Oxford University Press, Oxford, 2013, p 46.
24. Ibid p 76
25. Wikipedia, Focal plane tomography, https://en.wikipedia.org/wiki/Focal_plane_tomography, last accessed 18/07/21.
26. M. A. Zapala, K. Livingston, A. S. Phelps and J. D. MacKenzie, Digital tomosynthesis of the pediatric elbow, *Pediatr Radiol.* 49(12), Nov 2019, 1643–1651. doi: 10.1007/s00247-019-04444-y. Epub 2019 Nov 4. PMID: 31686170.
27. S. Bates, E. Beckmann, A. Thomas and R. Waltham, *Godfrey Hounsfield: Intuitive Genius of CT.* British Institute of Radiology, London, 2012, p 10.
28. David Whitfield, The story of how Sir Peter Mansfield invented the MRI scanner and won the Nobel Prize, *Nottingham Post*, 12/03/19, https://www.nottinghampost.com/news/health/story-how-sir-peter-mansfield-2635109, last accessed 18/07/21.
29. S. Bates, E. Beckmann, A. Thomas and R. Waltham, *Godfrey Hounsfield: Intuitive Genius of CT*, British Institute of Radiology, London, 2012, p 78.
30. Ibid p 60
31. Michael Frayn, *The Human Touch: Our Part in the Creation of a Universe*, Faber and Faber, London, 2006, p 32.
32. 1917 is a movie that feels like a videogame – in a good way, Wired, 01/03/20, https://www.wired.com/story/1917-videogame-movie/, last accessed 18/07/21.
33. Chris Taylor, '1917' is half movie, half video game, all genius, *Mashable*, 10/12/19, https://mashable.com/article/1917-movie-review/?europe=true, last accessed 18/07/21.
34. Todd Martens, From '1917' to, yes, 'Parasite,' video games are even influencing prestige movies, *Los Angeles Times*, 11/02/20, https://www.latimes.com/entertainment-arts/story/2020-02-11/oscars-1917-parasite-video-games-movies, last accessed 18/07/21.
35. Mark Cousins, *Women Make Film*, 2019, https://www.womenmakefilm.net/
36. Tad Leckman, Shapeshifting films, *Sanctuary Moon,* https://tadleckman.wordpress.com/2012/10/29/shapeshifiting-films/, last accessed 18/07/21.
37. *The Outer Limits (1963–1965)*, Internet Movie Database, https://www.imdb.com/title/tt0056777/characters/nm0674775, last accessed 18/07/21.
38. From *On Photography* by Copyright © Susan Sontag 1971, 1974, 1977, published by Farrar, Straus and Giroux 1977, Allen Lane 1978, Penguin Books 1979, 2019. Reproduced by permission of Penguin Books Ltd. ©
39. L. Cartwright, *Screening the Body: Tracing Medicine's Visual Culture*, University of Minnesota Press, Minnesota, 1995, p 142.
40. R. Porter, *The Greatest Benefit to Mankind: A Medical History of Humanity from Antiquity to the Present*, HarperCollins, London, 1997 p 649.
41. A. Patrizio and D. Kemp (eds) *Anatomy Acts: How We Come to Know Ourselves*, Birlinn, Edinburgh, 2006.
42. Ibid -Jonathan Sawday, The Paradoxes of Interiority, p 2
43. S. Bartsch, *The Aeneid - A New Translation Vergil*, Profile Books, London, 2020.
44. G. Wood, Aleksander Ceferin's 'long nose' issue highlights the VAR offside conundrum, *The Guardian*, 8 December 2019, https://www.theguardian.com/football/blog/2019/dec/08/aleksander-ceferin-long-nose-var-offside-conundrum, last accessed 18/07/21.

6 Similes, Similarities, and Symbolism

Label Don Diego, after Velasquez. Oil on linen by Derrick Guild. Courtesy of the artist and the Scottish Gallery

DOI: 10.1201/9780367855567-6

A little while ago I went litter picking along a beautiful stretch of coast near North Berwick. Actively on the lookout for bits of rubbish, it was astonishing how many seashore creatures and bits of naturally occurring beach debris were able to mimic the appearances of various items of plastic waste. White shells resembled fragments of polystyrene cups, crab legs pass as cigarette butts, blanched seaweed ribbon masquerades as plastic tape, and razor clams double for both plastic straws and (if the other way around) shards of brown glass. Bird poo, it turns out to my frustration, can resemble pretty much any item of litter in the right light or vantage point.

We saw in Chapter 2 how our visual system is inclined to actively impose meaning on the world we survey. While remarkably accurate in most situations, the experience of mistaking one thing for another is a universal and familiar experience. Mistaken identity forms the bedrock of numerous comedy plays throughout history but has also resulted in countless tragic miscarriages of justice. In the natural world, the propensity of animals to confuse the visual appearances of different entities is widely exploited in the form of camouflage and mimicry. It seems I am not the only one to have been misled by my beachcombing nemesis. Several animals have evolved an external appearance that strongly resembles a bird dropping as a means of camouflage, including the bird dung crab spider (Figure 6.1), and the swallowtail caterpillar (Figure 6.2).

Back on the streets of Edinburgh, splats of bird poo on the pavement can resemble things other than litter. Take a look at Figure 6.3, a photograph of a bird dropping, alongside Figure 6.4 the closest line drawing I could find to match the striding figure I perceived. Much like Rorschach ("ink blot") tests or spotting shapes in the clouds, the appearances deciphered may reveal more about the viewer than the image. Perhaps my subconscious self was encouraging me to stride on and stop wasting time looking at bird droppings. A few years prior to this I was again looking down at Edinburgh pavements, this time peering at chewing gum. Artist Juliana Capes had created a temporary work, *Earthly Bodies*, for the 2017 Edinburgh Art Festival consisting of drawing chalk lines between embedded bits of chewing gum on the paving slabs, such that they resembled diagrams of stellar constellations.[1] This seemingly simple manoeuvre succeeded in transforming these usually unpleasant reminders of antisocial behaviour into substitute celestial bodies and elevating them from the gutter into the stars.

Our ancestors have, of course, been gazing at actual stars for millennia and spotting all sorts of resemblances. Either the most underwhelming or alternatively most imaginative established constellation is *Canis Minor*, which can be seen in the second quadrant of the northern hemisphere sky, consisting of just two named stars, *Gomeisa* and *Procyon*, and purported to resemble one of Orion's hunting dogs.[2] More recent advances in astronomy have enabled visualisation of distant galaxies and nebulae, many of which have been named on the basis of their visual similarity to familiar earthly structures. These include the Horsehead Nebula (Figure 6.6), the Crab Nebula (Figure 6.7), and the Sombrero Galaxy (Figure 6.8).

The appearance of the distinctive Mexican hat has not only been used in relation to astronomical phenomenon. In March 2020, UK Prime Minister Boris Johnson stated his intention to "squash this sombrero", referring to the anticipated spike in

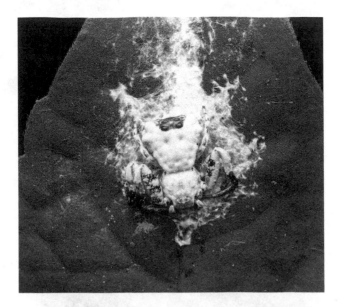

FIGURE 6.1   Photograph of the bird dung crab spider. Image credit: Alen thien / Shutterstock

FIGURE 6.2   Photograph of a caterpillar of the Giant Swallowtail butterfly, found in the rainforest in Costa Rica. Its appearance mimics that of a bird dropping to avoid predation. Image credit: Brian Magnier /Shutterstock

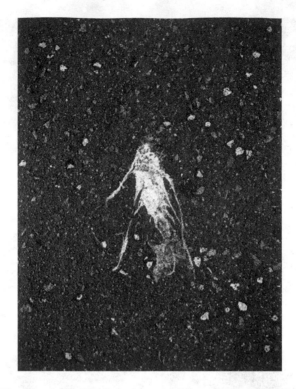

FIGURE 6.3   Photograph of a bird dropping on a pavement, Edinburgh. Image provided by the author.

FIGURE 6.4   Line drawing of a woman walking. Image credit: 1494 /Shutterstock

FIGURE 6.5 Photograph of the M83 galaxy. Messier 83, Southern Pinwheel Galaxy, M83 or NGC 5236 is a barred spiral galaxy in the constellation Hydra. Image credit: NASA Images. /Shutterstock.com

FIGURE 6.6 Photograph of the Horsehead nebula. Image credit: McCarthy's PhotoWorks/ Shutterstock

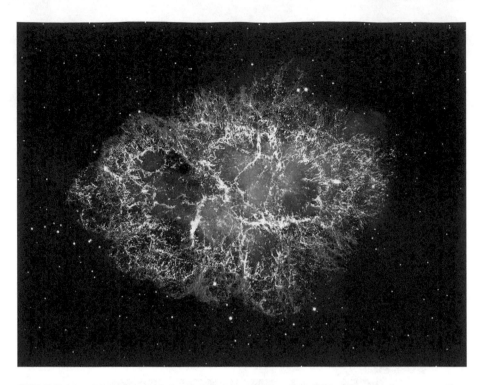

FIGURE 6.7   Photograph of the Crab nebula, in the constellation Taurus. Supernova Core pulsar neutron star. Image credit: Allexxandar/ Shutterstock

FIGURE 6.8   Photograph of the Sombrero galaxy, Messier 104 (M104). Image credit: Mohamed Elkhamisy/ Shutterstock

COVID cases represented in graph form during a Downing Street press briefing.[3] Widespread use of metaphor was to follow from both politicians and journalists in describing the coronavirus pandemic, utilising words like "battlefield", "frontline", "weapons", and "arsenal". The use of military metaphor and analogy was flagged up as problematic contemporaneously,[4] and the longstanding use of war-based language in relation to health and disease has been criticised for many years.[5] Recent research has suggested such bellicose metaphors in the context of cancer are counterproductive at best and at worst detrimental to both care and prevention.[6]

These are legitimate concerns, but our day-to-day vernacular is so rife with metaphors, similes, and analogies that it becomes difficult to imagine communicating without them. Indeed, language is often described as a collection of faded metaphors, with most words – if traced back sufficiently far – originating in some sort of metaphor or analogy.[7] Being in the business of describing appearances and providing names or labels for diseases on the basis of these appearances, the challenge of delivering understandable reports without setting foot in a metaphorical minefield (so to speak) is particularly salient within radiology.

NAMING WITHOUT SHAMING

As every parent, and most likely every pet owner, will know, the business of naming things can be a tricky business. Sharing my own name with the infamous King of Pop, I'm well aware of the consequences of the act of naming. I should perhaps make it clear that I was not named after the pop star, even if he had already achieved a significant level of celebrity by the time of my birth. While I must admit to occasionally adding "like the King of Pop" when asked for my name over the phone, this name sharing has, at best, been something of a mixed blessing. I discovered a few years ago that an abstract I had submitted for a national scientific meeting was thrown in the bin by the organisers of the conference who thought it was a wind-up on account of my name. As a general rule, it is usually myself on the receiving end of name-related wind-ups rather than delivering them.

Historically, a large number of diseases have been named after specific individuals, most commonly the person who first described that condition. Such eponymous disease labels have pros and cons. For example, "Perthes' disease" is certainly a snappier label than "idiopathic avascular osteonecrosis of the upper femoral epiphysis", and the former is the most commonly used term for this pathology in the UK. But in North America, this entity is called "Legg–Calvé–Perthes disease", acknowledging two other physicians involved in describing the condition, and dropping the genitive apostrophe following the view that this suggested possession of the condition.[8] Determining without controversy whether an individual deserves "ownership" of a particular disease entity is increasingly fraught. The disease now known as "Granulomatosis with Polyangitis (Wegener's)", was previously called Wegener's granulomatosis but Dr Friedrich Wegener was not the first physician to describe this condition, and the use of the eponymous terms for several related vasculitic diseases was deemed confusing and outdated. Revelations concerning Wegener's involvement in the Nazi regime in the 1930s and 40s accelerated the rebranding, announced in

2011, although with the recommendation that his name remained in parentheses for several years to smooth the name change.[9]

Eponymous labels can cause other difficulties. One individual may have more than one disease named after them (Paget's disease may refer to unrelated disorders of the bone, breast, penis, or vulva). Two individuals may have two diseases named after them, but share the same name; Kennedy's disease (aka spinal and bulbar muscular atrophy), named after William R. Kennedy, is not the same as Kennedy's syndrome, (a constellation of findings found in frontal lobe brain tumour patients, named after Robert Foster Kennedy). The name of the disease may relate to the individual who first described the condition, a famous individual who suffered from the disease (Lou Gehrig's disease) or a fictional character sharing some characteristics of the disease (Alice in Wonderland syndrome, Havisham syndrome, Munchausen syndrome).[10]

Non-eponymous terms run into other difficulties. Acronyms can be confusing (ASD; autistic spectrum disorder, or atrial septal defect? CP, cerebral palsy or child protection? etc.), implied natural history may be misleading (so-called "non-ossifying fibromas" do indeed, with time, ossify), and some names are just plain bewildering ("pseudopseudohypoparathyroidism"). Geneticists may have to counsel parents that their child suffers from a debilitating disease caused by a gene labelled "Sonic Hedgehog", or suffers from a disorder named after the resemblance of the child's facial features to make-up worn in traditional Japanese theatre ("Kabuki Syndrome").[11]

At the time of writing, NASA is in the process of "examining its use of unofficial terminology for cosmic objects as part of its commitment to diversity, equity and inclusion".[12] Examples of terms requiring attention include nebula NGC 2392, known as the "Eskimo Nebula", and a pair of spiral galaxies (NGC 4567 and NGC 4568) previously known as the "Siamese Twins Galaxy". In cases such as these, where the nicknames "have historical or cultural connotations that are objectionable or unwelcoming" NASA will only use the official, International Astronomical Union designations.[12] This, if a little unimaginative, is most likely the safest option to avoid future controversy. The tendency to allocate boring names to astronomical entities was parodied in Antoine de Saint-Exupéry's *The Little Prince* numerous decades prior to the current undertaking,[13] but the seemingly mundane label "M83" allocated to a spiral galaxy (Figure 6.5) has become the name of a successful French synth-pop outfit,[14] so the merits of any name are always open to discussion. Berlin metro stations,[15] Greene King pubs[16] and independent record labels[17] are likewise in the process of acquiring updated names to replace offensive historical names.

Conditions or characteristics which have been associated with prejudice or stigma have proved difficult to label without the label itself becoming pejorative. Linguists have coined the phrase "euphemism treadmill" for terms which regularly require replacement on account of the outdated label having become an offensive term.[18]

So there may be potentially heated debate about what something should be named, but there may also be confusion as to what that particular thing is and how it should be classified. In the UK there was a famous legal case to determine, for tax-related purposes, whether a Jaffa Cake was indeed a cake or actually a biscuit.[19] This was echoed more recently by the Irish courts' finding that the wheat-based substrate within which sandwich fillings were deposited in Subway outlets could not be classified as bread.[20] Lockdown restrictions in the UK also recently raised the question of whether a scotch egg could legitimately be counted as "a substantial meal".[21] These serve as reminders that it is not necessarily the label itself that is important, but the act of categorisation and the reasons behind it that require consideration.

As we have already seen, within medicine the business of naming diseases (*nomenclature*) and classifying them (*nosology*) can be very challenging. Within radiology and medical imaging, the intertwined nature of verbal and visual language further complicates this endeavour. Radiologists attempt to use words with precision in their reports, but leaving aside the business of diagnostic labels for now, even apparently uncontroversial descriptive terms can become problematic. I discourage trainees from using the word "sinister" in their reports (for example "no sinister bone lesion seen") partly on the basis of the word's pejorative origins related to left-handedness, partly due to Hammer horror or haunted house movie type associations. I suggest they use "concerning" as a replacement. One of my colleagues objects to the use of "unremarkable" in reports (which I admit to using frequently) on the basis that it is inherently paradoxical to state something is unremarkable, having, in the process, just remarked up on it. There is ongoing debate as to whether "appearances are consistent with ..." conveys more or less certainty than "appearances are compatible with ..." and on and on it goes.

THE PROBLEM OF NORMALITY

Most radiologists would nevertheless agree that one word – despite being universally used, and universally understood – can also be the hardest to include in reports. For politicians "sorry" was always said to be the hardest word to say (although in recent years "I'm sorry if ..." seems to have become easier to say than "I'm sorry for ..."). Amongst radiologists "normal" is typically the hardest word to commit to in a radiology report. The fear that a subtle abnormality hidden away in a radiograph or scan will come back to bite the radiologist who reported the examination (or worse still harm the patient) is endemic throughout X-ray departments. Despite this, countless imaging examinations are labelled normal on a minute-by-minute basis, and any examination labelled as abnormal is clearly reliant on a notional normal appearance from which there has been a deviation. The notion of normality, however, deserves a little more scrutiny.

The concept of an "ideal" human form has been prevalent throughout human history in both art and anatomy. Figures 6.9, 6.10, 6.11, and 6.12 are studies by Da Vinci, Michelangelo, Dürer, and Hogarth each demonstrating a preoccupation with idealised proportions in human anatomy. In *The End of Average* Todd Rose examines some of the difficulties of ideal, normal, and average body concepts.[22] A more

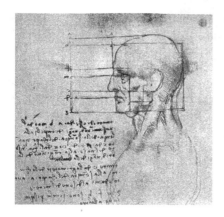

FIGURE 6.9 Sketch of a human head in profile, with linear horizontal lines located at the levels of particular facial features to calibrate anatomical proportions. Credit: Leonardo da Vinci, the anatomist: (1452–1519)/by J. Playfair McMurrich. Attribution 4.0 International (CC BY 4.0).

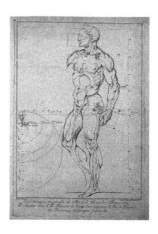

FIGURE 6.10 Sketch of a male figure in lateral view with anatomical proportion lines superimposed. Adapted from an original sketch by Michelangelo Buonarroti. Credit: History and bibliography of anatomic illustration/by Ludwig Choulant; translated and annotated by Mortimer Frank; further essays by Fielding H. Garrison, Mortimer Frank, Edward C. Streeter; with a new historical essay by Charles Singer, and a bibliography of Mortimer Frank, by J. Christian Bay. Wellcome Collection. Attribution 4.0 International (CC BY 4.0).

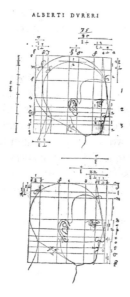

FIGURE 6.11 Outlines of two human heads in lateral profile with superimposed horizontal proportion lines. By Albrecht Dürer, 1557. Credit: De symmetria partium humanorum corporum libri quatuor, a germanica lingua in latinam versi / [Albrecht Dürer]. Wellcome Collection. Attribution 4.0 International (CC BY 4.0).

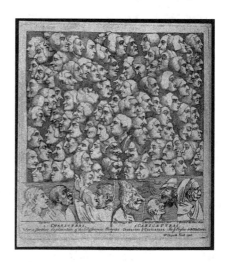

FIGURE 6.12 Characters and caricatures; the idealised heads of Saint John and Saint Paul are contrasted with grotesque heads, above them rises a cloud of faces showing different expressions. Etching by W. Hogarth. Credit: Wellcome Collection. Attribution 4.0 International (CC BY 4.0).

recent manifestation of idealised body morphology highlighted by Rose is a statue known as "Norma" sculpted by Abram Belskie on the instruction of gynaecologist Robert L. Dickinson, using averaged size measurements from some 15,000 young women. A contest was held in 1945 in which competitors were invited to submit body measurements matching those of "Norma". None of the 3,864 entrants (including the winner) matched the nine specified dimensions.[23] Likewise, summarising research conducted by Gilbert S. Daniels in the US Air Force in the 1950s, Rose points out that of 4,063 pilots, not one fitted within average limits on ten measured body dimensions, concluding that a cockpit designed to fit an average pilot will, in fact, fit nobody.[24] The variations in body morphology from one individual to another means we need to tread carefully when utilising an average version, as we saw in relation to brain architecture and atlases in Chapter 4. The notorious phrase "Normal for Norfolk", abbreviated to "NFN" in questionable medical correspondence of the 20th century and more recently repurposed as a title of a BBC television programme,[25] reminds us that we need to consider both the particular population being considered and the intent of those assigning the label of normality.

Yet surely if we remove the subjective and troublesome aspects of beauty and aesthetics, and put variations in size to one side, it is not unreasonable to study the structure of the human body to determine the "normal" topographical relationships of all the bits and pieces that make up the body? This is the fundamental basis of how medical science works. The basics of what structures and arrangements make a "normal" human body would seem to be pretty uncontroversial, but when you get into the details things become a bit trickier.

The Human Genome Project offers some insight. While undoubtedly a spectacular milestone in the history of genetics, I would take issue with it being labelled using the definite article. "A Human Genome" would in some respects be more appropriate, given that the original project consisted of analysing a solitary human genome. Although this was composed in total from around 30 individuals to cover all 46 chromosomes in their entirety, the large majority of the DNA used was acquired from only around half a dozen people, with the bulk of the genetic code being derived from a solitary male African-American donor from upstate New York.[26] The limitation of the project at the time of its completion in 2001 was that it had sequenced the genetic code representative of only one (admittedly slightly composite) person out of a population of around five billion (eight billion at the time of writing). This genome could therefore not legitimately represent all of humanity by a long stretch. It is the differences between individuals' DNA that interest geneticists, and making sense of inherited genetic disease, tumour mutations and all the other ambitions stated by The Human Genome Project was always going to require analysis of genes and genomes from a large number of subjects, not just one man from New York.

Long before embarking on the Human Genome Project, geneticists knew that variability within the gene pool is a good, healthy thing. Restriction of gene pool/consanguinity leads to potentially fatal inherited genetic disease. While an increasing number of inherited genetic diseases can be attributed to specific mutated genes causing loss of function or similar, the complexity of the interplay between genetics and environmental factors means that, more broadly, it is very challenging to label particular variant genes as "good" or "bad". If a eugenicist wanted to produce a single "perfect" human genome, the practical impossibility of such a thing ought to stop them in their tracks even if the flawed logic of such an undertaking didn't.

So trying to determine an ideal human form starting with the molecular building blocks is – at best – problematic. As a radiologist I spend my working life looking at body structures a long way downstream from the gene coding level. The bones I look at may have been subject to environmental factors – perhaps deformed from a previous fracture or showing "growth arrest lines" from periods of illness. They may be of reduced density from disuse or lack of vitamin D, or conversely abnormally dense from exposure to heavy metals such as lead. Genetic factors may result in variable morphology of particular bones, but this is usually a complex polygenic process. The genetic mechanisms responsible for osteogenesis imperfecta are better understood, known to be caused by as many as 33 separate mutations in the gene that codes for Type 1A collagen.[27] But even in this context, the characteristic radiographic appearances of osteogenesis imperfecta – multiple bone fractures from relatively trivial trauma – rely on an interaction with the environment. The body at the macroscopic level that I am used to viewing it, therefore, represents an interplay between the individual's unique genetic code and all the environmental factors they have been exposed to since conception. As anatomist and broadcaster Prof Alice Roberts puts it:

> Nature and nurture, genes and environment are inseparable. Culture is woven into biology in a way that can never actually be unravelled, it is not just draped over the top.[28]

Although when I look at a radiograph of a bone to see if it is broken, I have a notional idea of what a normal, say, femur looks like, trying to conjure a specific image in my mind is a bit more of a challenge. Certainly at the level of the trabecular patterning visible within the bone (the fine areas of linearity seen centrally within each bone) all bets are off, with the variability similar to that seen in an individual's fingerprints. But a template of approximate femoral architecture is certainly utilised all the same. In making a comparison of the image of a unique, particular, and specific human subject in front of me with this notional "normal" bone, I am invoking a highly abstract conceptual device. Constructing a "population-level" morphological version of the femur, I have embarked on what I suppose you might call the Human Thigh Bone project.

THE GOLDEN COUPLE: TOO MUCH INFORMATION?

The problem of one individual acting as a representation of all humanity that we encountered in relation to the Human Genome Project has some resonance with the human figures engraved on the Pioneer plaques. The Pioneer 10 and 11 spacecraft, launched in 1972 and 1973 respectively, were the first man-made objects to achieve escape velocity from the Solar System.[29] Housed within each probe is a 9 by 6-inch rectangular plaque composed of gold anodised aluminium engraved with information about Earth, just in case alien life should ever intercept it. Two naked human figures are portrayed, one male, one female, shown in Figure 6.13. The style is pretty minimalist, consisting of linear, almost sketch style figures. Yet the figures are recognisably white/Caucasian and the hairstyles are rather reminiscent of the 1970s era. It can certainly be argued that these figures – on ethnicity grounds alone – are not representative of the human race in its entirety. Although the figures were drawn by a woman (Linda Sagan) critics suggest it is sexist to have the man raising a hand in

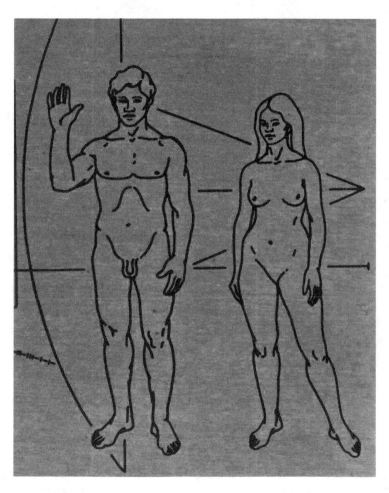

FIGURE 6.13   Cropped segment of the Pioneer Plaque, showing two human figures, one male, one female. The male figure's hand is raised in a gesture of greeting. Credit: NASA Ames.

an active gesture of greeting, while the woman stands in an indifferent posture.[30] In addition, we have the man standing to the right of the woman, reflecting longstanding artistic convention (in turn based on religious doctrine) of right dominance – we could accurately describe the woman's position as sinister. If fig leaves were present, we would instantly label these figures as Adam and Eve – compare the composition to Figure 6.14.

Just a little detail in the image has meant the first representation of humanity to potentially fall into alien hands (or equivalent) is sexist, racist, biased towards

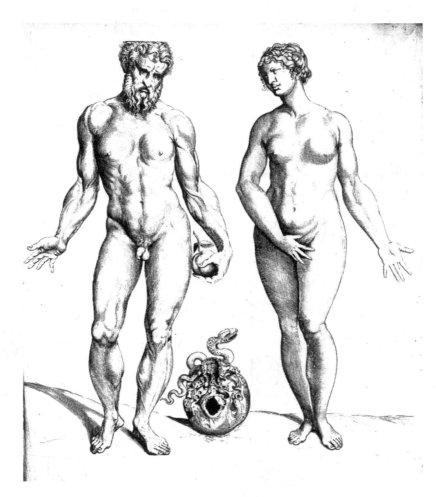

FIGURE 6.14   Adam and Eve style figures from Renaissance-era anatomical textbook. The male figure holds an apple in his left hand and a snake is present on a skull located between the feet of the two figures. Credit: Compendiosa totius anatomiae delineatio aere exarata / [Thomas Geminus]. Wellcome Collection. Public Domain Mark.

Judaeo-Christian iconography and with questionable hairstyles. The absence of pubic hair and external female genitalia is another issue altogether. The figures caused controversy at the time, although largely in relation to their nudity.[30] On launching the Voyager 1 space probe in 1977 (and now the most distant man-made object from Earth) the figures were altered to silhouettes to sidestep further accusations of obscenity and racial bias, and with both figures hands kept down by their sides.[31] The man still appears to be located to the right of the woman, although as we have seen previously, without the benefit of side markers in a silhouette image, it is possible we could be seeing the figures viewed from the front or from behind.

To consider another situation in which too much detail in an image can be problematic, have a look at Figures 6.15, 6.16, and 6.17. Each of these images is a representation of a black pepper plant; a woodcut from the 16th century (Figure 6.15), a line engraving from the 18th century (Figure 6.16) and a photograph from the 21st century (Figure 6.17). To modern eyes we are inclined to view the photograph as the definitive, most accurate image, with the line drawing claiming the silver medal and the woodcut illustration a distant third. Yet if the purpose of each image is considered, we may need to re-evaluate. For botanists seeking to illustrate characteristic features of a plant species, rather than a likeness of one plant, in particular, the issue is of some importance. A named species consists of a group of individual organisms with a degree of variability within their morphology and genetics, so getting to grips with the criteria used to define one species from another is also inherently problematic. The level of detail offered by a photograph likewise becomes a challenge. A little nibble out of one of the leaves, some partial decay, etc. – these are features which may be present in individual members of the species but are not a characteristic feature of the species as a whole. We need a photo-shopped version of the plant, in which such imperfections have been airbrushed out, and perhaps Figure 6.16 offers us that – a pepper plant rendered in sufficient detail to allow identification, but with idiosyncratic variations such as leaf bites omitted. Botanical illustrations also have the benefit of allowing the entire plant to be in focus and without shadows obscuring particular areas.[32] Perhaps even then an overly-literal field explorer might assume that the fruit branches need to be pointing upwards as shown in Figure 6.16, and disregard plants in which the pepper pods dangle downwards. The highly stylised representation offered in Figure 6.15 is not quite as inaccurate as we might first suppose. Provided the devices used in the construction of the image are understood by the viewer, it may offer a highly efficient means of communicating the key features of a pepper plant, in the same tradition as the map-like representation of the brachial plexus we considered in Chapter 4 (Figure 4.14). An understanding that the image is a generalised representation becomes crucial to interpreting the image, and the resemblance to a pepper plant (idealised or not) is conditional upon this contextualisation.

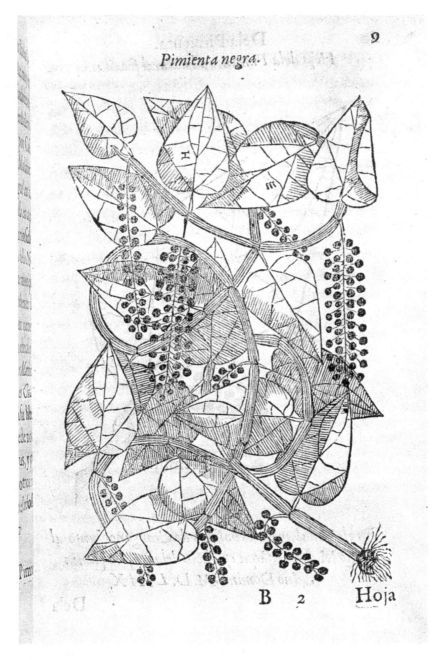

FIGURE 6.15 A woodcut of a pepper plant, from the 16th century. Credit: Tractado de las drogas, y medicinas de las Indias orientales. Wellcome Collection. Attribution 4.0 International (CC BY 4.0).

PLANTE D'UN POIVRIER.

FIGURE 6.16   A line engraving of a pepper plant from the 18th century (Fig 6.16). Pepper plant (Piper nigrum L.): fruiting stem. Line engraving after C. de Bruin, 1706. Date [1718]. Credit: Wellcome Collection. Attribution 4.0 International (CC BY 4.0).

FIGURE 6.17   A photograph of a black pepper plant showing plant with green berries and leaves. Note portions of the leaves are out of focus and that areas of damage or decay are visible in the leaves. Image credit: C. Aphirak/ Shutterstock

CLASSIFICATION SYSTEMS/LINGUISTIC RELATIVITY

The difficulty of images representing a species (too much detail becomes overly prescriptive; too little prevents identification at all) is mirrored in disease classification systems; if too general it is useless, too detailed it becomes unmanageable. There is a reason that most film reviews are accompanied by a rating out of five stars, rather than out of ten or a percentage score, and it is probably no coincidence that many of the most successful classification systems used in medicine to determine disease subtypes, fracture patterns or traumatic injuries to organs also rely on a five-way split.

In nosology the concept of a spectrum of manifestations or severity is a useful means of describing a wide variation of phenotypes and genotypes.[27] "On the spectrum" has become a widespread, if not always appropriate, shorthand to describe behaviours and tendencies associated with Autism. In recent decades there has been increasing recognition that characteristics such as gender and sexuality are better described by a more nuanced classification than the traditional binary categorisations. UK newspaper reports of 71 gender identity options available on Facebook implied this was far too many,[33] but the addition of a question on gender identity in the UK 2021 census reflects growing awareness that a simple two-way split is insufficient to capture the complexity of gender.[34] Whatever it is that is being classified will always be constrained by the categories available.

In the UK the DVLA requires vehicles to be allocated into one of the following 20 "standard" colours: beige, *black*, *blue*, bronze, *brown*, buff, cream, gold, *green*, *grey*, ivory, maroon, *orange*, *pink*, *purple*, *red*, silver, turquoise, *white*, and *yellow*.[35] We find ourselves in a typical classification conundrum – clearly it would be possible to list more colour options, but the larger the number of options the more unmanageable it would become. While I can't see the logic of having four choices (beige, cream, ivory, and white) to cover off-white type hues, but only two (green or turquoise) to cover everything green, perhaps as a UK motorist I should just count myself lucky there are 20 options rather than 5. Indeed, linguists argue that only 11 "basic colour terms" are required for English speakers to be able to agree on the colour of any given object (these are highlighted in italics in the list above).[36] Some languages expand on these 11 (Russian has separate terms for light and dark blue, Hungarian has separate terms for light and dark red) but many other languages have fewer terms, with a large number of languages using a single word to describe both green and blue (referred to by linguists as a "grue" term). Examples include Vietnamese *xanh*, Thai *khiaw*, Chinese *qīng*, Korean *pureuda*, and Japanese *ao*. Some languages (most common in West African and Native American languages) combine red and yellow under a single descriptive word (a "rellow" term). Studies suggest there is accompanying variation in colour perception according to which language an individual speaks, with, for example, reduced ability to discriminate different shades of blue and green amongst those using "grue" terms.[37, 38, 39]

Linguists since the 19th century have suggested that the language an individual speaks may have a wider impact on how they think and perceive the world more broadly, an idea often termed linguistic relativity or relativism.[40] The theory has often been named after 20th-century linguists Edward Sapir and Benjamin Lee Whorf, known as either the Sapir-Whorf hypothesis or Whorfian hypothesis, but not without controversy.[41] The interplay between culture and neurophysiology in this setting has not been fully untangled, and the significance of such differences has been largely down-played by contemporary linguists.[40] Nevertheless, it seems there is a connection of some sort between the linguistic terms available to an individual to describe the colour and the manner in which colour is perceived. It is certainly difficult to deny that if a word to describe a particular entity does not exist in a given language, a speaker may struggle to describe it. We saw in Chapter 4 that the Guguyimidjir language of aboriginal Australians has no words for "left" and "right". In *Through the Language Glass*, Guy Deutscher points out that a number of languages use a single term to describe "hand" and "arm" (such as *yad* in Hebrew), and in Hawaiian "arm", "hand", and "finger" are all included under the same term.[42] It is not only the eye of the beholder that influences the visualisation of the world around us but also the dictionary they keep on their bookshelf.

LETTERS AND LOGOS, SYMBOLS AND SIGNS

Figure 6.18 shows an illustration of the archaeological site by the mountain of Serabit el Khadim, in the Sinai Peninsula of Egypt. This mid-19th-century lithograph predates the discovery (by William and Hilda Petrie in 1904) of inscriptions at a temple within the site, now thought to be the world's first alphabet. Known as the proto-Sinaitic scripts, the inscriptions are composed of modified Egyptian hieroglyphs, the earliest of which date from around 1850 BCE.[43] These are believed to have been carved by migrant workers from Canaan, working in the nearby Turquoise mines and represent the transformation of pictographic written text into an alphabet-based language.[44] Previous written languages, including Egyptian hieroglyphs, utilised the Rebus principle in which words unrelated to the original glyphs could be constructed on a "sounds like" basis (e.g. in English the word "belief" could be constructed using a picture of a bee followed by a picture of a leaf). By contrast in the proto-Sinaitic scripts, individual glyphs (or characters) were used not to represent a specific entity (or the sound of the corresponding word), but instead a very short characteristic sound (or phoneme), such that by combining several glyphs in sequence words could be formed, entirely unrelated to the pictographic meaning of the original glyphs. This also enabled new words to be created more readily, without the necessity of devising new glyphs.[44, 45]

The system, devised by a group of illiterate migrant workers, is thought to be a direct precursor of both the Ancient South Arabian script and the Phoenician alphabet, the latter of which produced the written forms of a large number of modern languages, including the letters from which you are reading this sentence.[43]

The letter "A" for example can be traced back to the Egyptian hieroglyph of an ox, which was stylised and simplified in the proto-Sinaitic script to 𐤀 then further simplified and rotated in the Phoenician alphabet to 𐤊 then rotated again and adjusted a little in the archaic Greek alphabet to **A**. Turning an "A" upside down, we can still see a resemblance to a triangular Ox's head and two horns sticking out.[43, 44]

Written language, traced back to its origins, is inseparable from a pictographic visual representation of the world. The popularity of emojis in direct messaging provides a more contemporary reminder of this.

Now cast your eye over the multiple iterations of arrows shown in Figure 6.19. These are variants of the symbol for recycling. Several of these are immediately

FIGURE 6.18 Lithograph/aquatint showing the archaeological site by the mountain of Serabit el Khadim, in the Sinai Peninsula of Egypt. by Louis Haghe after David Roberts, 1849. Credit: Wellcome Collection. Attribution 4.0 International (CC BY 4.0).

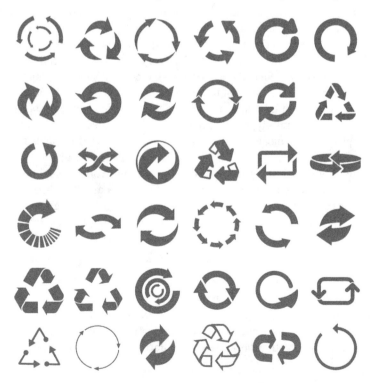

FIGURE 6.19 Iterations of arrows associated with logos for recycling. Image credit: Pavlo S/ Shutterstock

recognisable to me from food packaging or recycling bins, but others – whilst largely similar in design – do not have that familiarity or conceptual link. In some cases the design is more readily linked in my mind to another symbol or logo; a few of the single arrow designs, particularly the one in the left-hand column with the dashed tail, make me think of icons from some computers telling me the program is still loading; the rectangular designs are more suggestive to me of a circuit diagram from an electronics textbook. Some also remind me of cyclical processes described in medical textbooks such as the Krebs cycle in biochemistry, or the "plan-do-study-act" cycle approach to healthcare quality improvement projects, and another reminds me of the logo for the Chicago Cubs baseball team. Collectively, looking at all these arrows has also reminded me of the opening credit sequences to Alfred Hitchcock's *North by Northwest* (1959) and Steven Spielberg's *Catch me if you can* (2002). The context of interpreting such symbols is all important – I would probably make the recycling connection for nearly all of these variants if any of them were printed on the side of a drinks can. Taken out of the intended context it is instructive to see how many associations these relatively simple designs can trigger.

One of the two arrow circular designs is reminiscent of the logo of Opel, the car manufacturer. I am not aware of a radiological sign named after the Opel logo, but there is certainly a Mercedes-Benz sign and a Honda sign, both named in relation to resemblance to the distinctive logos of these car companies. We will consider these very shortly but while thinking about large multinational vehicle manufacturers, a brief detour to consider Peugeot is instructive. In *Sapiens*, historian Yuval Noah Harari uses the French car company to illustrate how shared ideas can transcend any basis in physical reality. In 1896 – the year the first clinical radiographs were acquired – Armand Peugeot set up the motor company bearing his name. Harai likens the legal procedures and rituals required by French law to bring the Peugeot company into being to the ceremonial process of a Catholic mass culminating in the transubstantiation of bread and wine into the flesh and blood of God. Pointing out the resemblance of the Peugeot logo to the Stadel lion man, a prehistoric ivory sculpture composed of a lion's head on a human-like body, Harari suggests Peugeot is similarly a concoction of collective human imagination but – along with other corporations, nation states, and religions – an extremely powerful one.[46]

Radiological signs are also a product of human imagination, if not quite so world-changing. These are characteristic appearances of particular disease processes or manifestations as shown on imaging investigations which resemble an unrelated object or entity. For a radiological sign to be successful/useful the resemblance needs to be specific to one disease process and the comparison reference sufficiently well recognised to be widely understood. Such signs have borrowed heavily from familiar or distinctive appearances found in a variety of settings[47]:

- Botanical: Pine cone bladder, Ginkgo leaf sign, Cloverleaf skull, Tulip bulb sign, Tree-in-bud pattern, Drooping lily, Bamboo spine

- The animal kingdom: Raccoon eyes sign, Eye of the tiger sign, Moose head appearance, Hummingbird sign, Brahma bull sign, Staghorn renal calculus, Bird's beak or rat-tail oesophagus, Scottie dog
- Food: Pretzel sign, Coca cola bottle sign, Rice grain calcification, Popcorn calcification, Onion skin periosteal reaction, the Linguine sign, Hot cross bun sign, Cottage loaf sign, Coffee-bean sign
- Clothing: Inverted Napoleon hat sign, Corduroy sign of vertebral haemangioma, Bow tie appearance cervical spine, Boot-shaped heart

The use of familiar objects for these purposes has echoes of anatomical nomenclature. The scaphoid and navicular bones are named after their boat-like appearance (using the Greek and Latin roots, respectively), the cube-like cuneiform bones are named after dice, and the coccyx is named after a cuckoo's beak. In each of these cases there is an approximation required on the part of the individual – the bones resemble these things up to a point, but not all that closely and we may wonder how specific the comparison is supposed to be in any case. As with my Human Thigh Bone project, we require a notional reference standard – not a particular boat, set of dice or cuckoo, but a Platonic archetype embodying the morphology being conjured in each case.

However, there is a convention of exclusivity assigned to some radiological signs and associated descriptive terms. A "dripping candlewax" appearance is – by radiological convention – used to describe the characteristic appearance of a rare bone condition called melorheostosis (illustrated in Figure 6.20), so you will not find this term in radiology reports used to describe another bone condition, hereditary multiple exostosis (shown in Figure 6.21), despite a definite resemblance (see Figure 6.22).

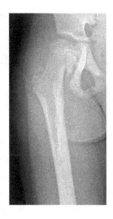 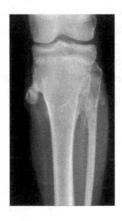 

FIGURE 6.20 Cropped radiograph of the upper femur showing features of melorheostosis. Image provided by the author.

FIGURE 6.21 Cropped radiograph of the upper tibia showing features of hereditary multiple exostosis. Image provided by the author.

FIGURE 6.22 Photograph of a candle, with dripping candlewax running down the side. Image provided by the author.

A very particular frame of reference is also required in some cases – the "champagne glass" appearance of the pelvis described in achondroplasia refers to the old fashioned coupe rather than the narrow flutes currently in vogue.

Back to car logos then. The Mercedes-Benz sign refers to either a triradiate pattern of gas sometimes seen in gallstones on abdominal radiographs or the cross-sectional appearance of the aorta on CT in the context of aortic dissection, again producing the distinctive triradiate shape. More recently the "inverted Mercedes-Benz sign" has been described in relation to spinal subdural haematoma as seen in axial plane on MRI – so much for exclusivity. Figure 6.23 shows an example of the "Honda sign" – a sacral insufficiency fracture shown on a nuclear medicine bone scan – the characteristic distribution of uptake of radiotracer in the pelvic bones resembling the H shape of the motor company logo, shown in Figure 6.24. The company presumably has the precise dimensions and angles of the "H" specified in detail for copyright purposes, but in the context of radiology interpretation the likeness need only be approximate to prompt recognition of the pathology.

The plasticity of the letter H demonstrated in this context is also reflected in the near endless variations of text characters available whether via typefaces in mechanical print or the idiosyncrasies of an individual's handwriting. When taking my French GCSE as a 16-year old, I struck upon what I considered to be a cunning ruse for the written part of the exam. For nouns where I was unable to remember the gender (and thus unsure whether to put "le" or "la" ahead of the word) I would write "l" followed by an ambiguously figured letter halfway between an "e" and an "a". My hope was that the examiner would "see what they want to see". I have no way of knowing if this worked, but I did get a good grade, so who knows. There is ongoing debate in Germany related to the use of gender-specific nouns in legal documents, with concern raised that this may compromise gender equality, alongside a broader discussion of whether such linguistic terms are outdated.[48] Perhaps what I have labelled as an examination ruse was actually a helpful solution to such challenges.

The phenomenon of skipping spelling errors whilst reading is well recognised and research suggests that we are all adept at extracting meaning from text even when bearing only a faint resemblance to the original intended version.[49] Personalised car registration plates which don't quite spell out what we see at first glance attest to this, and a similar phenomenon is also demonstrated by the names of variations in the SARS-CoV-2 virus. A change located at position 484 resulting in lysine (designated as K) being expressed instead of glutamic acid (designated as E) was described in the seemingly dreary code "E484K", and tyrosine (Y) replacing asparagine (N) at position 501 was assigned "N501Y" – neutral scientific terms that sidestep the emotive issue of the locations at which such variants are thought to have arisen, much like the boring but inoffensive galaxy names. However, virologists have nicknamed E484K as "Eeek!" and N501Y as "Nelly", on the basis of a relatively tenuous resemblance to these words. These variations in the genetic code relate to the Spike protein, likened to a flick knife, also maintaining the weaponry metaphor theme.[50]

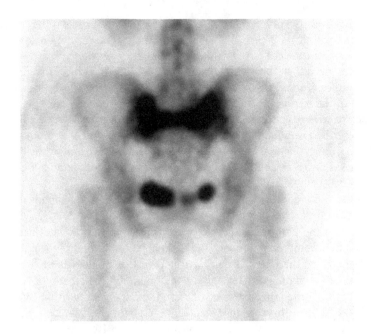

FIGURE 6.23 An example of the "Honda sign" – a sacral insufficiency fracture shown on a nuclear medicine bone scan – the characteristic distribution of uptake of radiotracer in the pelvic bones resembling the H shape of the motor company logo. Image courtesy of Dr D Patel.

FIGURE 6.24 Image of the Honda logo. Image credit: tanuha2001/ Shutterstock

JOINING THE DOTS

Whether looking up at the stars or down at chewing gum (or bird droppings) on the pavement, we are familiar with joining the dots to impose a pattern, structure, or meaning regardless of whether it is there. *Pareidolia* is the term used by psychologists to describe this phenomenon which periodically manifests itself in such forms as Buzz Lightyear seen in a carrot, religious figures appearing in burnt toast, and both cats and buildings that resemble Adolf Hitler.[51] From the International Space Station, astronaut Chris Hadfield spotted a whale gulping down krill (appropriately enough the *mouth* of the Humber, in Hull), an elephant (the geographic contours of the bay of Naples), and a bird pecking at a seed (the Crimean coastline).[52] As a radiologist I am not immune to this phenomenon and have also projected some bird-like appearances onto various radiology examinations (see Figures 6.25–6.30). I recall a rather endearing example of radiological pareidolia occurring while I was doing an ultrasound on a little boy, perhaps three years old or so. Watching the screen as I scanned his kidneys, he was convinced he saw a monster, asking "what's a monster doing in my tummy?" and then, apparently unfazed by the idea of a monster lurking within, started to do perfect monster "grrr!!!!"ing accompanied by classic monster claw movements.

Going right back to the earliest cave art, it has been suggested that it was this tendency to see resemblances that initiated representational art. In his commentary *De Statua*, Renaissance polymath Leon Battista Alberti stated:

> I believe that the arts that aim at imitating the creations of nature originated in the following way: in a tree trunk, a lump of earth, or in some other thing were accidentally discovered one day certain contours that needed only a very slight change to look strikingly like some natural object.[53]

Alberti's suggestion has been supported by more recent art historians such as E.H. Gombrich and by Martin Gayford who in *A History of Pictures* speculates that a pebble embedded in a cave wall provides the starting point for a lion's eye, with other contours of the cave wall resembling ears or shoulders – a pre-existing template onto which some further embellishments produce the whole figure of a lion.[54]

Leonardo da Vinci was certainly open to the creative possibilities of pareidolia:

> Look at certain walls stained with damp, or at stones of uneven colour ... you will be able to see in these the likeness of divine landscapes, adorned with mountains, ruins, rocks, woods ... expressions of faces and clothes and an infinity of things. [55]

In 1952, mathematician Alan Turing, best known for breaking the Enigma code during World War II and his insight into the potential of modern computing, published a paper "The chemical basis of morphogenesis" in the *Philosophical Transactions of the Royal Society*.[56] Taking inspiration from the work of D'Arcy Thomson's 1917 book *On Growth and Form* in which mathematical principles are used to explain diverse biological phenomena, Turing proposed a mechanism to explain how the symmetrical, spherical ball of cells of an early embryo starts to develop asymmetry (tackling the L-R problem we touched on in Chapter 4) and downstream complex

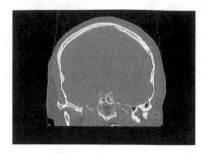

FIGURE 6.25 Coronal CT reconstruction showing the skull base, resembling an owl. Image provided by the author.

FIGURE 6.26 Line drawing of an owl. Image credit: Robert Varga/ Shutterstock

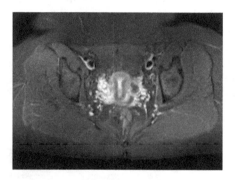

FIGURE 6.27 MRI scan of the pelvis, in which appearances resemble the head of an eagle. Image provided by the author.

FIGURE 6.28 Photograph of a bald eagle. Image credit: Wilfred Marissen/ Shutterstock.

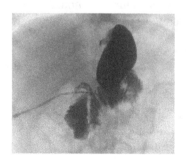

FIGURE 6.29 Fluoroscopic examination in which barium in the stomach resembles a vulture. Image provided by the author.

FIGURE 6.30 Photograph of a Lappetfaced Vulture, Torgos tracheliotus, perched in a tree. Image credit: pixelpics/ Shutterstock

morphology. The mathematical model described the interaction of two antagonistic chemicals (or "morphogens") labelled as an activator and an inhibitor.[57]

While preparing the paper, Turing is said to have shown diagrams from his modelling to his bemused University of Manchester colleagues asking if they saw a resemblance to the patches of colour on a cow's coat.[58] Regardless of what his colleagues thought, the resemblance that Turing perceived was more significant than simple pareidolia. What have now been named "Turing Patterns" are found not only in cow print dappling, leopard's spots, and zebra stripes, but also in the repeating ridges and ripples in sand dunes (the interaction between the wind and the ridges following the same activator/inhibitor mechanisms).[59] While the embryological processes have yet to be fully characterised, there is growing evidence to suggest the mechanisms Turing described are involved.[57] On occasion, spotting a resemblance may reveal a deeper significance.

With all of this dot-joining in mind, have a read of the following paragraph:

> *"From the area of the head of the litres removed millimols and there is 11 ulnar rather than is of normal R P L breasts and pain. Grain pain. A follow-up plain film. The patient was on MRI"*

This is what the voice recognition software I use at work came up with when I played it the gobbledygook/yodelling parts of "Hocus Pocus" by 1970s Dutch prog rock band *Focus*. Now, admittedly I have only succeeded in turning one form of gobbledygook into another, but the final sentence is actually a sentence, and the penultimate sentence sounds like it might belong in a radiology report (perhaps more commonly followed by "is suggested" or "is advised"). It certainly feels like it would take less time for yodelling played into voice recognition software to produce a passable radiology report than those infinite monkeys set to work on typewriters. This is not altogether surprising, as the software is specifically designed to transform audio signal into predetermined English words and phrases, employing probabilistic "best guesses" to allocate those which are most likely to be used in a radiology report. Hence largely random sound (OK, I cheated a bit by using recognisably human voices) is skewed into something beginning to resemble a radiology report. This exercise reverses the more common scenario of perfectly enunciated reports as dictated by the radiologist being turned into nonsense by the VR software.

Now take a look at Figure 6.31, a cropped segment of a drawing of a tree I can see across the street from my living room window. I did this sketch during the winter, the leafless branches silhouetted against the sky. Using a little artistic licence I used some dots or discontinuous short lines to represent the smallest branches/twigs. Figure 6.32 shows the same photograph after applying an imaging "filter" readily available as a smartphone application. The algorithm has literally joined the dots to help smooth out some of the discontinuous lines, taking some more "best guesses" as to which dotted trajectories should be represented as continuous lines.

In radiology, some modified "joining the dots" software is utilised in a variety of settings including 3D reconstructions of CT angiograms to demonstrate blood vessels with the surrounding soft tissues removed, as well as the virtual bronchoscopies and colonoscopies we encountered in Chapter 5. More specifically, let us consider

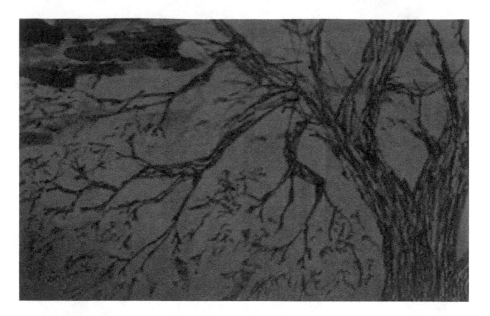

FIGURE 6.31 A cropped segment of a drawing of a tree in winter. Image provided by the author.

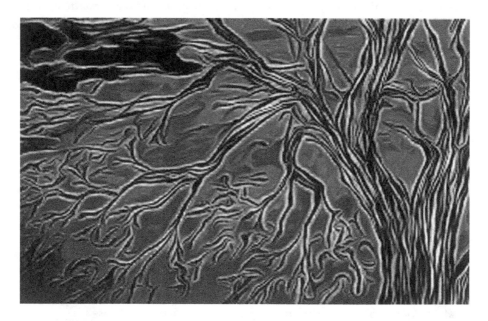

FIGURE 6.32 The same photograph as Figure 6.31 after applying an imaging "filter" readily available as a smartphone application. Image provided by the author.

an application now in widespread use in neuroimaging. Diffusion Tensor Imaging (DTI) is an MRI-based tool that allows visualisation of white matter pathways in the brain and spinal cord. The technique utilises information about the movement of water molecules within each voxel, determining the net direction in which water is moving at each site (see Figure 6.33). By applying a "best-fit" algorithm, the voxel-based vectors are joined up to produce lines, and lines following a common trajectory are clumped together to represent white matter pathways – bundles of neuronal axons connecting one part of the brain to another. DTI now has a number of well-established diagnostic uses and can assist neurosurgeons planning complex operations such as the removal of a brain tumour with the minimum of deficit to functional neural tissue. It also produces spectacular images of the internal structure of the brain such as Figure 6.35, and Figure 4.13 from Chapter 4. Yet when we look at such images we need to remain aware that the image is constructed to see the brain in a very particular way, rather than how it is in reality.

Tractography images of the brain certainly seem to add credence to the "wiring" analogy of the brain – we see what looks very much like wires distributed through the brain and it is natural to make the comparison to a computer, even if computers with this many visible wires have not been produced for several decades. However, diffusion tensor imaging has also been utilised in other body regions including the kidneys and heart (see Figure 6.36), producing similar wire-like appearances. In these situations, we are seeing an analogue of the renal architecture and alignment of muscle fibres within the heart respectively, and would therefore not talk about the wiring of the kidney or heart.

I have juxtaposed *Variations* by Paul Klee (Figure 6.34) alongside the DTI map (Figure 6.33). The resemblance is largely coincidental, but provides a relevant digression. Klee is known to have held naming parties, in which he invited friends to suggest titles for as yet untitled paintings.[60] E.H. Gombrich, John Berger, and others have emphasised the role an artwork's title may have on its interpretation, particularly in more abstract works, where the viewer may require some pointers. A telling example of the title or caption informing the interpretation provided by Gombrich is *The Nuremberg Chronicle* (1493), in which an identical woodcut illustration of a medieval city is used several times, using a different city name (including Damascus, Milan, and Mantua) as the annotation on each occasion.[61]

Whether it is projecting the essence of a city onto identical buildings or turning uniform filaments into neurons, nephrons, or muscle fibres according to whether the label informs us we are looking at a brain, kidney, or heart, we are highly adept at seeing what we want to see and imposing a context-dependent interpretation as we do so. In the specific situations that such Diffusion Tensor Images are used this is entirely legitimate. We can make an informed inference that the linear structures shown in the brain do indeed correspond to white matter tracts on the basis that such tracts have been demonstrated through the dissection and histopathology of countless brains of deceased individuals over hundreds of years. Radiologists and neurosurgeons alike are aware that the "wires" shown in a tractography image are only a macroscopic approximation of neurophysiological reality, albeit a potentially very useful one.

Taking such images out of context, however, can risk a reductive, potentially dehumanising attitude, intended or otherwise – if we visualise people as machines,

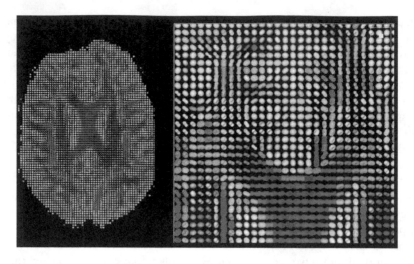

FIGURE 6.33 The left panel is an axial image of the brain showing diffusion vectors as discrete ellipsoids, corresponding to the net direction of water molecule movement at each voxel. The right panel shows a magnified view of the region around the genu of the corpus callosum. Different colours allocated to each ellipsoid indicate the direction of diffusion, in this context surmised to indicate the direction of white matter tracts. Image courtesy of Prof J.R. Alger. The Diffusion Tensor Imaging Toolbox. *Journal of Neuroscience* 30 May 2012, 32 (22) 7418-7428; DOI: https://doi.org/10.1523/JNEUROSCI.4687-11.2012

FIGURE 6.34   Variations by Paul Klee. Image credit: Everett Collection/ Shutterstock

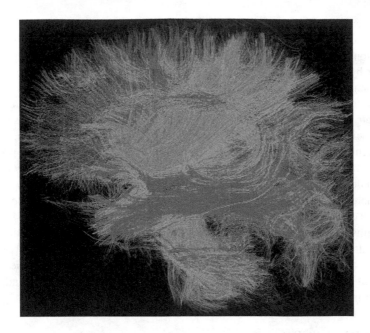

FIGURE 6.35   Diffusion tensor imaging showing white matter tracts within the brain of a neonate. As in Figure 6.33 different colours indicate directional information about individual filaments. Image courtesy of Dr M Bastin, University of Edinburgh

FIGURE 6.36   This image shows the swirling arrangement of cardiac fibres in the left ventricle using diffusion tensor imaging. Credit: Heartstrings. University of Oxford, Dr P. Hales/BBSRC. Attribution 4.0 International (CC BY 4.0).

perhaps we start treating them as such. Tractography has made at least one notable appearance in popular culture – an image of DTI white matter tracts from the Human Connectome Project adorning the cover of platinum-selling album *The 2nd Law*, released by rock band Muse in 2012. Both the *New Musical Express* and *Rolling Stone* speak of brain "circuits" in their description of the artwork,[62, 63] reinforcing a mechanistic view of the body. Putting the reductive dimensions aside, there is ongoing debate amongst neuroscientists as to whether the computer analogy is a helpful one in trying to understand how the brain works. In *The Idea of the Brain*, Matthew Cobb highlights some of the flaws of the computer/brain metaphor.[64] He points out that the linearity of coding steps occurring one after the other implied by the computer analogy and likened to a domino rally, fails to convey the highly interconnected nature of neural networks, a dimension also not well captured in the "A to B" trajectory of the wire-like filaments in tractography images. It is certainly important to recognise the limitations of what imaging can reveal about the function of the brain as distinct from its structure. In *Neuropolis: A Brain Science Survival Guide*, comedian and writer Robert Newman deftly satirises the lofty claims made by some functional MRI research papers such as those claiming to visualise what individuals are dreaming, alongside some wry takedowns of mechanistic metaphor.[65]

METAPHORS AND MOTIFS

In the highly compelling and heartbreaking documentary *Once upon a time in Iraq* (dir. James Bluemel, 2020) eye witness accounts of the war and its aftermath are provided by a wide range of participants including photographer Ashley Gilbertson. He recalls his frustration when people compared his work to movies or described his photographs as "cinematic", viewing this as diminishing the colossal human suffering we hear about throughout the documentary. However, it is interesting to compare this perspective with one of the US soldiers featured in the same documentary, who in describing coming under fire during night combat directly likens the experience to X-wing fighters being shot at by laser cannons and Tie-fighters while attacking the Death Star in *Star Wars* (dir. George Lucas, 1977). Our visual points of reference are not limitless, and are heavily influenced by the culture we are immersed within – including cinematic references, as you might have noticed reading this book. While writing this chapter Matt Hancock, the UK government Health Secretary, was revealed to have used the movie *Contagion* (dir. Steven Soderbergh, 2011) as a shorthand for conveying the importance of vaccine production in tackling the pandemic.[66]

Western art has always relied on a set of readily identifiable figures and motifs, drawing upon religion and mythology. The iconography of Christian martyrs relies heavily upon the means by which particular Saints were executed (or in St Sebastian's case as we have seen, attempted execution). The classical tradition of Greek and Roman gods and mythological figures also includes characteristic visual motifs or totems to assist easy recognition. The mythology of the Star Wars saga follows in these footsteps, which I grew up with. The next generation may be more familiar with *The Avengers* as a set of readily identifiable heroes, although the inclusion of Thor provides a link to ancient Nordic mythology. Superman, who we will return to shortly, can also stake a claim as an instantly recognisable visual icon.

In the previous chapter we saw how well-established visual motifs or tropes can provide a familiar framework on which to project in the form of St Sebastian. Much of art history can be summarised as the process by which new generations of artists build or modify pre-existing visual templates, as E.H. Gombrich argues in *Art and Illusion*.

In *Every Painter Paints Himself*, Simon Abrahams contends that in portraits of other individuals artists have a tendency to paint themselves as much as their subject.[67] An impressive array of portraits juxtaposed alongside self-portraits by numerous artists from the Renaissance era onwards, does at first look very persuasive.[68] One artist for whom the evidence is overwhelming is Rembrandt van Rijn. The Dutch master was a prolific self-portraitist (with between 86 and 98 extant works of himself documented) but also included his likeness in numerous other works.[69, 70] A paper highlighting this aspect of self-identification in Rembrandt's work targeted specifically at radiologists encourages a similar approach in the reporting of imaging examinations – imagine it is yourself in the scanner and you might scrutinise the study that little bit more carefully.[71]

However, I'm not convinced that *every* painter paints themself. Looking through lots of these examples, there are recurring collective similarities – most of the faces (whether subject or artist) are angled obliquely to the viewer (such that those holding our gaze look at us out of the corner of their eyes), mouths are closed, and left upper lighting prevails. Era-dependent features, such as beards, hairstyles, and ruffs contribute to the resemblances. If we take a look back at the four self-portraits I assembled in Chapter 4 there appear to be resemblances between Rembrandt and Dürer, and between van Gogh and Bernard. It seems that the collective influence of artistic convention, combined with the *écriture* of the specific artist is as much a contributor to these resemblances as a psychological urge to represent oneself.

ANATOMICAL MOTIFS

Take a look at Figure 6.37, a study of a human cadaver by Albrecht Dürer. While on a technical level it is difficult to fault, the subject matter strikes me as a bit of a mess – bones missing, tattered ragged bits of flesh, skin irregularly torn. While there are recognisable bones and anatomical structures, it is not clear where our focus should lie. Unlike more familiar representations of human anatomy in the Renaissance era this looks like a carcass that has been picked over by scavengers, not the result of painstaking dissection.

Compare this to the (approximately contemporaneous) anatomical figures seen in Figures 6.38–6.41. These are illustrations from Andreas Vesalius' game-changing *De humani corporis fabrica* published in 1543. While the body is depicted in various states of dissection, the human frame remains intact insofar as it is a whole body (i.e. head-to-toe) rather than discrete body regions. The figures also maintain life-like posture as though still alive (the characteristic poses are also seen in various of the St Sebastian and anatomical figures featured in Chapter 5, and the *écorché* figure with the chopped-off arm in Chapter 3). A key feature is the specific nature of the bodily structures being demonstrated – the muscular system, the vascular system, the skeleton. Each illustration invites us to focus on these particular structures whilst other extraneous structures are either concealed by skin or muscle, or have disappeared altogether. A similar, if less polished approach to anatomical representation had been

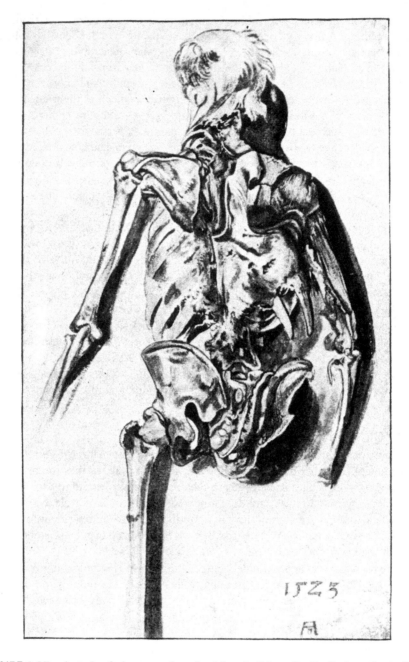

FIGURE 6.37   A study of a human cadaver by Albrecht Dürer. Credit: Skeleton Study Sheet, by Dürer. Wellcome Collection. Attribution 4.0 International (CC BY 4.0).

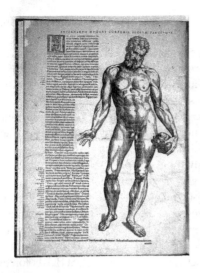

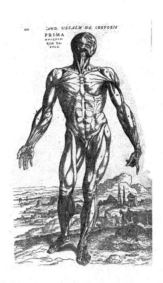

FIGURE 6.38 Nude male anatomical figure, holding a skull in his left hand. Illustration from de humani corporis fabrica by Andreas Vesalius. Credit: Andreae Vesalii Suorum de humani corporis fabrica librorum epitome. Wellcome Collection. Public Domain Mark.

FIGURE 6.39 Male anatomical écorché figure. Illustration from de humani corporis fabrica by Andreas Vesalius. Note the identical characteristic pose as seen in Figure 6.38. Credit: Andreae Vesalii Suorum de humani corporis fabrica librorum epitome. Wellcome Collection. Public Domain Mark.

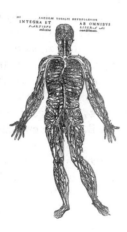

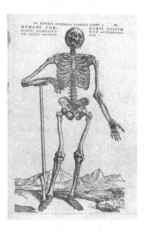

FIGURE 6.40 Whole-body anatomical figure, showing the vascular system in isolation. Illustration from de humani corporis fabrica by Andreas Vesalius. The characteristic pose seen in Figure 6.38 and Figure 6.39 is maintained. Credit: Andreae Vesalii Suorum de humani corporis fabrica librorum epitome. Wellcome Collection. Public Domain Mark.

FIGURE 6.41 Whole-body skeleton depicted in a characteristic pose, but with right arm flexed and leaning on a spade. Illustration from de humani corporis fabrica by Andreas Vesalius. Credit: Andreae Vesalii Bruxellensis, scholae medicorum Patauinae professoris De humani corporis fabrica libri septem. Wellcome Collection. Public Domain Mark

in use for at least 200 years prior, shown in Figures 6.42–6.45. Outside of anatomy textbooks, these whole body system motifs have become firmly established within the popular visual imagination, with a modern take on such archetypes shown in Figure 6.46. The Disney animated feature *Inner Workings* (dir. Leonardo Matsuda, 2016), borrows heavily on these tropes. Over the course of a day the hero finds himself governed by the agenda of half a dozen bodily systems including the central nervous system, circulatory system, respiratory system, gastrointestinal tract, and the urinary tract, not always working in a coordinated fashion. The respective depictions of each of these systems, again shown in whole body format, not only draws upon anatomical motifs of centuries duration, but is also strongly reminiscent of the surface rendered 3D reconstructions of CT scans, such as those shown in Figure 6.47.

In art, anatomy, and radiology there is a strong relationship between *what* is being portrayed and *how* it is depicted, such that the representation of particular structures is governed by a rulebook of conventions. Such conventions are well demonstrated in the depiction of muscular anatomy. If the previous chapter laid out a case for St Sebastian being the patron saint of both the anatomical plane and anatomical label, St Bartholomew can make a claim for patron of the anatomical *écorché* figure, and again demonstrates the intersection of religion, culture, art, and anatomical science. Bartholomew was martyred by being flayed and typical depictions portray the saint holding his skin or draping it around his body, often holding a flaying knife as well. Figure 6.48 shows the sculpture *St Bartholomew Flayed*, by Marco d'Agrate (1562). Skin is draped like a cloak, and the underlying muscular anatomy graphically revealed. This statue is housed within Milan Cathedral, and while clearly an embodiment of Renaissance anatomical studies, is primarily a devotional work. By contrast, Figures 6.49 and 6.50 are taken from renaissance era anatomy books, yet the influence of Bartholomew's iconography is keenly felt.

St Bartholomew has probably had less impact on contemporary visual culture compared to St Sebastian, but Damien Hirst's 2006 sculpture *Exquisite Pain* is a flayed *écorché* figure directly modelled on Renaissance-era depictions of the saint, and several of his later works including Verity (2012, shown in Figure 6.51) show partially flayed or semi-dissected figures. The video accompanying Robbie Williams hit single *Rock DJ* (2000) culminates in the pop star delivering a striptease in which not only his clothes but also his skin is discarded, leaving a Bartholomew-like muscle figure engaging in some very unsaintly dancing.

Once the skin is removed muscles are revealed, but in the Bartholomew-modelled *écorché* figures only those muscles immediately beneath the surface are visible, and not in their entirety. To allow individual muscles to be displayed Vesalius employed "exploded" muscle dissections in which muscles are unhitched from their insertion sites (at least at one end), an example of which is shown in Figure 6.52. A modified version of this technique has been used in several of Gunther von Hagens' "plastinated" human figures in his Bodyworlds exhibitions, an example of which is shown in Figure 6.53. In *The Transparent Body* Jose van Dijck draws a comparison of such "expanded bodies" to Umberto Boccioni's sculpture *Unique Forms of Continuity in Space* (1913),[72] shown in Figure 6.54. While Boccioni's work may have been exploring movement and dynamism as much as anatomy, the conceptual link to Vesalius is very plausible. Likewise, the muscle-like structures featured in the Transformers robots in various recent movie iterations (such as that shown in Figure 6.55) would

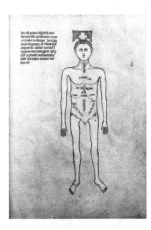

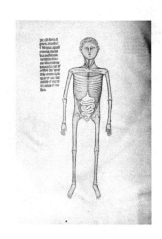

FIGURE 6.42 Anatomical illustration from the 14th century showing a whole-body male figure. Credit: Guido de Vigevano, miniature anatomical figures, 1345. Wellcome Collection. Attribution 4.0 International (CC BY 4.0).

FIGURE 6.43 Anatomical illustration from the 14th century showing a whole-body male figure, predominantly skeletal in appearance but with loops of intestines seen in the abdomen. Credit: Guido de Vigevano, miniature anatomical figures, 1345. Wellcome Collection. Attribution 4.0 International (CC BY 4.0).

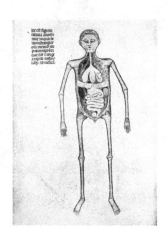

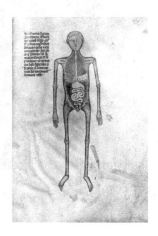

FIGURE 6.44 Anatomical illustration from the 14th century showing a whole-body male figure. The limbs are skeletal in appearance but with loops of intestines seen in the abdomen, and lungs and blood vessels shown in the chest. Credit: Guido de Vigevano, miniature anatomical figures, 1345. Wellcome Collection. Attribution 4.0 International (CC BY 4.0).

FIGURE 6.45 Anatomical illustration from the 14th century showing a whole-body male figure. Appearances are similar to Figure 6.44, but with the lungs and blood vessels removed from the chest, allowing the oesophagus to be seen connecting to the stomach. Credit: Guido de Vigevano, miniature anatomical figures, 1345. Wellcome Collection. Attribution 4.0 International (CC BY 4.0).

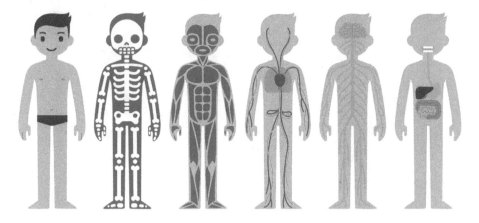

FIGURE 6.46   Cartoon-style anatomical figures, with characteristic representation of specific body systems: skeletal, muscular, circulatory, nervous and digestive systems. Image credit: Sudowoodo/ Shutterstock

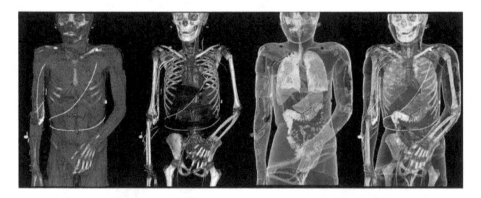

FIGURE 6.47   Surface rendered 3D reconstructions of CT scans. Image provided by the author.

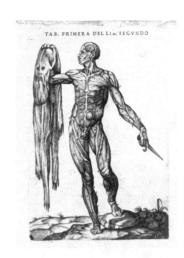

FIGURE 6.48   Photograph of the sculpture St Bartholomew Flayed, by Marco d'Agrate (1562). Image credit: Andrej Privizer/ Shutterstock

FIGURE 6.49   Flayed *écorché* figure, holding skin aloft in his right hand and a flaying knife in his left hand. Credit: Historia de la composicion del cuerpo humano. Wellcome Collection. Attribution 4.0 International (CC BY 4.0).

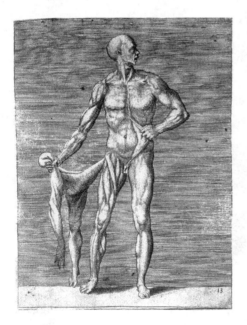

FIGURE 6.50   A male *écorché* figure, of whom only the right side of his body has been flayed. He holds the removed skin of his right side in his right hand and pulls back the skin of his left abdomen with his left hand. Engraving by G. Bonasone. Credit: Wellcome Collection. Attribution 4.0 International (CC BY 4.0).

FIGURE 6.51   Photograph of *Verity* (2012) by Damien Hirst. Image credit: A.G. Baxter/ Shutterstock

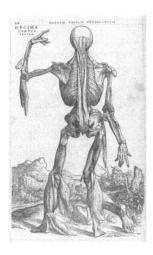

FIGURE 6.52   A male écorché figure demonstrating dissections in which muscles are unhitched from their insertion sites, falling away from the bones. Credit: Thirteenth muscle Tabula from Fabrica. Wellcome Collection. Attribution 4.0 International (CC BY 4.0).

FIGURE 6.53 Photograph of a plastinated human figure in a running pose, from Gunther von Hagens' Bodyworlds exhibition, demonstrating an "exploded" appearance of forearm muscles. Image credit: praszkiewicz/ Shutterstock

FIGURE 6.54 Photograph of Umberto Boccioni's sculpture Unique Forms of Continuity in Space (1913). Image credit: Everett Collection/ Shutterstock

FIGURE 6.55 Model of the robot Bumblebee displayed at Wulong National Park, China promoting a Transformers movie. The muscles-like strips of metal decorating the robot's limbs belong in the tradition of anatomical écorché figures. Image credit: GG6369/ Shutterstock

seem to owe more to such characteristic representations of human anatomy than to anything robotic.

KERNELS OF TRUTH?

In the Superman movie *Man of Steel* (dir. Zack Snyder, 2013) there are a couple of scenes involving X-ray vision. While most audience members will be unsurprised to discover that our hero possesses the ability to see through flesh and other solid structures, the film employs a less familiar introduction to this superpower. In a flashback scene to Clark Kent's childhood we see him as a young boy disturbed by the sight of his schoolteachers and classmates in various levels of transparency, resembling both the anatomical skeletons and *écorché* figures we have just been considering. Later in the movie the villain of the piece, General Zod, arrives on planet Earth to discover he, too, has X-ray vision and has a moment looking at his own outstretched hand which likewise varies in the level of transparency, the bones briefly revealed (like a radiograph) before flesh and glove fabric covers them over again. In each case the startled/disconcerted facial expressions combined with the rapidly shifting level of visual penetration suggests both characters are struggling to get to grips with this unfamiliar way of seeing the world, and in particular having difficulty fixing on a particular degree of transparency.

In the early days of radiography a similar period of adjustment was required to optimise exposure times and factors to obtain the best quality images. Figure 6.56 shows an example of an early hand radiograph with multiple exposure times utilised, increasing in duration from the fingertips to the wrist, demonstrating increasing penetration and visualisation of the bones as the exposure time is lengthened.

We saw in the previous chapter how modern cross-sectional imaging allows the body to be viewed from every conceivable angle. While tailored views and 3D reconstructions may be utilised for particular purposes, I suggested that radiologists tend to spend the most time looking in the axial plane, particularly for CT examinations. Modern CT scans also allow the radiologist to adjust the "windowing" of the images to optimise the greyscale display to the structures of most interest. The range of densities represented in any given slice can vary dramatically – from air within the lungs or bowel loops, to dense bone or perhaps metallic structures such as orthopaedic implants or the aneurysm coiling we saw in Figure 3.12. Accordingly, it is possible to display such images in a dizzying number of possible permutations. However, in regular reporting there are probably only around half a dozen preconfigured window levels commonly used, the most commonly used being soft tissue, bone, and lung. In certain situations, tailored adjustment of the windowing levels can be helpful to characterise a subtle abnormality, but the majority of a radiologist's time is spent looking at images using the common default settings.

In addition, CT image construction also utilises software to optimise the balance between image noise and spatial resolution for particular types of tissue, known as kernels (also called convolution algorithms or more simply "filters"). Bone kernels are of higher spatial resolution but at the cost of being noisier/grainier, while soft tissue kernels reduce the noise level but at the cost of spatial resolution. Once reviewing these separate datasets on a reporting station, the radiologist has a whole range

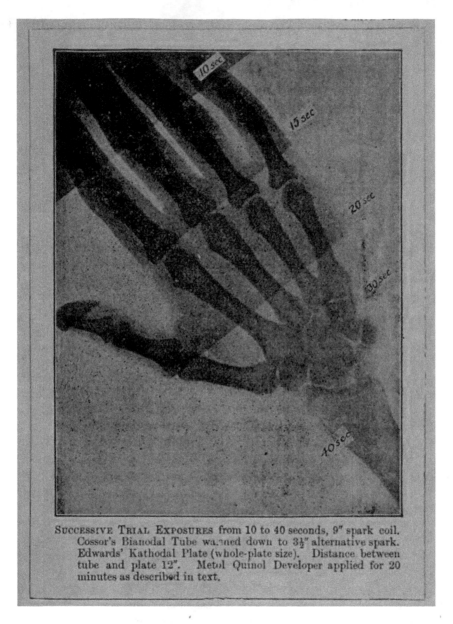

SUCCESSIVE TRIAL EXPOSURES from 10 to 40 seconds, 9″ spark coil.
Cossor's Bianodal Tube warmed down to 3½″ alternative spark.
Edwards' Kathodal Plate (whole-plate size). Distance between
tube and plate 12″. Metol Quinol Developer applied for 20
minutes as described in text.

FIGURE 6.56 An early hand radiograph with five different exposure times employed,
varying from 10 seconds over the fingertips to 40 seconds over the wrist bones. Illustration
from X-rays simply explained: *A Handbook on Röntgen Rays in Theory and Practice*/by R.P.
Howgrave-Graham. 1904. Credit: Wellcome Collection. Public Domain Mark.

of additional "post-processing" manipulations available, including the MPR and 3D reconstructions described in Chapter 5. Another manipulation that is closely related to convolution algorithms is the slice thickness used to view the images. Again, a large range of possible slice thickness options, but most radiologists will tend to polarise between "thin" slices (perhaps 1 mm or less) when looking at high contrast structures such as bones or lung structures, and "thick" slices (of around 4–5 mm) to look at most other soft tissue structures. I'm simplifying a little, and there are some additional tools that manipulate slice thickness along with other parameters, such as maximum (and minimum) intensity projection which are useful in evaluating the lungs. Nevertheless, the concept of radiologists narrowing potentially limitless imaging parameters down to a handful of most useful angles, window levels, and slice thicknesses still stands.

INTROMISSION/EXTRAMISSION

In ancient Greece, two theories of visual perception emerged. Democritus and Epicurus supported the idea of intromission which described objects and entities projecting resemblances of themselves known as eidola, which were captured by the eye. This model implied that the external world was composed exclusively of matter and phenomena that are perceived as complete, whole entities. By contrast, Plato believed in extramission, whereby rays are projected from the eyes to enable vision.[73, 74]

While both theories were ultimately disproved by an improved understanding of optics in the renaissance era, it is interesting that – at a metaphorical level – both ideas do provide some insight into the nature of visual perception. Pattern recognition described as template and prototype recognition by psychologists and the Gestalt model of perception rely on the mind constructing interpretations consisting of whole entities, not unlike the eidola of Democritus and Epicurus. However, in doing so there is an active dimension of projection, not dissimilar to Plato's eye rays, imposing a specific interpretation on the external world, originating from within the viewer.

The process of constructing visual appearances is not, however, confined to internal cognitive mechanisms. In shaping visual culture over hundreds and thousands of years, humans have become attuned to a certain way of seeing the world and more specifically characteristic representations of the body.

SUMMARY

We are deeply inclined to see resemblances. Language is integral to both the process of recognising such similarities and in providing a means of communicating them. In doing so we are reliant on the vocabulary that is available to us, and upon the visual milieu with which we are familiar, in the form of established motifs and visual tropes. The effect of a finite vocabulary and finite set of motifs is to both constrain the range of interpretations and descriptions of any visual encounter. Restrictions imposed by categories and classifications nevertheless also serve to maximise the efficiency of communicating the nature and meaning of visual information to others.

While the Whorfian hypothesis of linguistic relativism is not without its critics, the idea that the words we use to describe visual characteristics of the world can have a physiological impact on the way we do actually see the world adds a fascinating dimension. Just as cumulative life experiences are likely to impact an individual's worldview (in the political sense) it seems that cumulative linguistic and visual sensory input has a tangible effect on an individual's view of the world (in the physiological meaning). Culture and biology are inseparably intertwined.

The assessment of human anatomy and disease based on image-based techniques will, then, always provide an incomplete description. Yet we should not be too despondent. Have a look at two images of mouse anatomy both revealing a moment of analogue to digital conversion. Figure 6.57 shows an early image in the development of MRI – the first image of a mouse, performed by the pioneering Aberdeen team in March 1974. Each number denotes a cuboidal volume at a particular location (a voxel), and the value of the number denotes the intensity of signal from that particular site. The numbers correspond to pixels, but at this point in the process, the Aberdeen team were required to hand shade the squares using coloured pencils. The outline of the mouse is also hand-drawn, traversing individual pixels, much like the India ink line we zoomed in on in Chapter 3, again blurring the boundaries between the raw data and the imposed human interpretation of its meaning. The image captures a pivotal moment in medical imaging, a digitally acquired dataset rendered as a comprehensible image by a human hand.

Figure 6.58 shows the hand (or paw) of a mouse embryo at an early stage of development. The discrete, darkly dyed regions are areas in which particular genes have been expressed (possibly on the basis of the morphogenic principles described by Turing) which will initiate the process of apoptosis – cell death. By actively killing cells in these areas, the linear protrusions remaining will become fingers. The gaps between digits is an integral part of a functioning hand, and the gaps between disease categories are usually sufficiently small not to impair treatment. It is perhaps worth remembering that even things we consider to be a continuous spectrum – such as the wavelengths of electromagnetic radiation in sunlight – are not completely continuous on close inspection. Spectroscopic analysis of sunlight reveals characteristic parallel black lines (absorption spectra or Fraunhofer lines) corresponding to frequencies absorbed by specific elements. The barcode-like appearance of Fraunhofer lines reminds me of the (more evenly spaced) parallel strips of an anti-scatter grid, as mentioned in Chapter 3. In the case of the anti-scatter grid, the apparent loss of information in the form of undeviated X-rays that are absorbed by the metal strips is compensated by a net gain in information by filtering out unwanted noise from scattered X-rays. Likewise, the discontinuous framework that language (visual or verbal) imposes helps to streamline communication, eliminating (or at least reducing) noise and confusion. The Fraunhofer lines also bear a resemblance – and I promise this is the final resemblance of the chapter – to the tuning panel of an analogue radio set. I have emphasised how radiologists "tune in" to specific characteristic features of the imaging they are interpreting – looking at anatomy from particular angles or planes, making use of particular greyscale windowing, viewing images in a narrow range of slice thicknesses and utilising 3D reconstructions fashioned on longstanding anatomical archetypes.

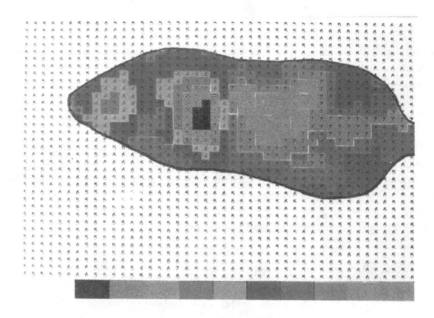

FIGURE 6.57   First MRI picture of a mouse, performed by the Aberdeen team in March 1974. Image © University of Aberdeen, reproduced with permission.

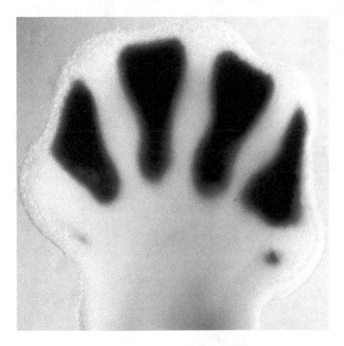

FIGURE 6.58   Light micrograph of a developing mouse limb showing gene expression in the hand plate. The regions highlighted will undergo apoptosis (cell death) to form the digits. Credit: Mouse hand plate. Kate Storey. Wellcome Collection. Attribution 4.0 International (CC BY 4.0).

The discrete, compartmentalised fashion in which we are required to view and describe both the body and the diseases which affect it is problematic on a philosophical level, but also highly effective on a pragmatic one. Even if the brain does not work like a computer, computers are nevertheless beginning to perform visual perception tasks based on the way we think the brain sees, utilising template matching pattern recognition alongside the joining-the-dot type algorithms we have encountered in this chapter. We will confront the challenges and opportunities of machine-based learning in the final chapter. Regardless of the framework within which they work, radiologists must reconcile the universal and the particular, ensuring that nuances of individual variability and evolving patterns of disease are neither lost in translation nor muddled in metaphor. To do so not only requires a grasp of anatomy, pathology, and imaging modalities of the day but also ongoing receptivity to new methods of visualising and describing the body – a willingness to tune into new radio stations as well as the tried-and-tested ones.

REFERENCES

1. Juliana Capes, Earthly bodies – Solo project and exhibition, Edinburgh Art Festival 2017– Writer's Museum, Edinburgh, http://www.julianacapes.co.uk/cv/, Last accessed 19/07/21.
2. Constellation Guide, https://www.constellation-guide.com/constellation-list/canis-minor-constellation/, last accessed 19/07/21.
3. Billy Perrigo, Coronavirus could hit the U.K. harder than any other European Country. Here's what went wrong, *Time*, https://time.com/5823382/britain-coronavirus-response/, last accessed 19/07/21.
4. D. Isaacs and A. Priesz, COVID 19 and the metaphor of war, *Journal of Paediatrics and Child Health* 57, 2021, 6–8. doi:10.1111/jpc.15164
5. S. Sonntag, *Illness as Metaphor*, Farrar, Strauss and Giroux, New York, NY, 1978.
6. D. Hauser, The war on prevention II: Battle metaphors undermine cancer treatment and prevention and do not increase vigilance, PsyArXiv, 4 Aug. 2019. doi:10.31234/osf.io/a6bvd
7. Stan Carey, Dictionary of faded metaphors, *Sentence First*, https://stancarey.wordpress.com/2011/05/04/dictionary-of-faded-metaphors, last accessed 19/07/21.
8. Wikipedia, Legg–Calvé–Perthes disease, https://en.wikipedia.org/wiki/Legg%E2%80%93Calv%C3%A9%E2%80%93Perthes_disease, last accessed 19/07/21.
9. Vasculitis Foundation, https://www.vasculitisfoundation.org/mcm_faq/what-is-granulomatosis-with-polyangiitiswegeners/, Last accessed 19/07/21.
10. Wikipedia, List of eponymous diseases, https://en.wikipedia.org/wiki/List_of_eponymous_diseases, last accessed 19/07/21.
11. National Organization for Rare Diseases, Kabuki syndrome, https://rarediseases.org/rare-diseases/kabuki-syndrome/, last accessed 19/07/21.
12. NASA, (Editor: Tricia Talbert): NASA to reexamine nicknames for cosmic objects, 05/08/20, https://www.nasa.gov/feature/nasa-to-reexamine-nicknames-for-cosmic-objects, Last accessed 19/07/21.
13. A. de Saint-Exupéry, *The Little Prince*, Egmont, London, 2009, pp 14–15.
14. Wikipedia, M83 (band), https://en.wikipedia.org/wiki/M83_(band), last accessed 19/07/21.
15. Vera Eckert, Mohrenstrasse: Berlin metro to change 'racist' station name by end of year, *Independent*, 04/07/20, https://www.independent.co.uk/news/world/europe/berlin-metro-mohrenstrasse-rename-racist-station-a9601641.html, last accessed 19/07/21.

16. Greene King, 14/01/21, https://www.greeneking.co.uk/newsroom/latest-news/green e-king-takes-anti-racist-stance-by-changing-names-of-four-pubs/, last accessed 19/07/21.
17. One Little Independent Records, https://www.olirecords.com/, last accessed 19/07/21.
18. Karen Stollznow, Ableist language and the euphemism treadmill, *Fifteen eighty four*, 11/08/20, http://www.cambridgeblog.org/2020/08/ableist-language-and-the-euphemis m-treadmill/, last accessed 19/07/21.
19. HM Revenue & Customs, *HMRC Internal Manual: VAT Food*, https://www.gov.uk/h mrc-internal-manuals/vat-food/vfood6260, Last accessed 19/07/21
20. BBC News, Subway rolls ruled too sugary to be bread in Ireland, *BBC News*, 01/10/20, https://www.bbc.co.uk/news/business-54370056, last accessed 19/07/21.
21. Archie Bland, Scotch egg is definitely a substantial meal, says Michael Gove, *Guardian*, 01/12/20, https://www.theguardian.com/world/2020/dec/01/scotch-egg-is-definitely-a-substantial-meal-says-michael-gove, last accessed 19/07/21.
22. T. Rose, *The End of Average: How to Succeed in a World that Values Sameness*, Penguin Books, London, 2015.
23. Ibid pp 5–7
24. Ibid p 4
25. Laurence Cawley and Jodie Smith, Normal for Norfolk: Where did the phrase come from? *BBC News*, 24/04/16, https://www.bbc.co.uk/news/uk-england-norfolk-3608 2307, last accessed 19/07/21.
26. Discovery: Human genome project's 20th Anniversary, Radio Broadcast, BBC World Service, 03/08/2020.
27. A. C. Offiah and C. M. Hall, The radiologic diagnosis of skeletal dysplasias: Past, present and future, *Pediatric Radiology*, 50(12), 2020, 1650–1657. doi:10.1007/s00247-019-04533-y
28. Prof. A. Roberts, Bodies, Episode 2, *BBC Radio 4 Broadcast* January 2021. Quoted with permission.
29. Mission Archives, NASA, 23/06/07 https://www.nasa.gov/centers/ames/missions/ar chive/pioneer10-11.html, last accessed 19/07/21.
30. Wikipedia, Pioneer plaque, https://en.wikipedia.org/wiki/Pioneer_plaque#Criticism, last accessed 19/07/21.
31. Wikipedia, Voyager 1, https://en.wikipedia.org/wiki/Voyager_1, last accessed 19/07/21.
32. E. T. Ben-Ari, Better than a thousand words: Botanical artists blend science and aes-thetics, *BioScience* 49(8), August 1999, 602–608, doi:10.2307/1313435
33. Rhiannon Williams, Facebook's 71 gender options come to UK users, *Telegraph*, 27/06/14, https://www.telegraph.co.uk/technology/facebook/10930654/Facebooks-71-gender-options-come-to-UK-users.html, last accessed 19/07/21.
34. Census 2021: Final guidance for the question "What is your sex?" Office for National Statistics, https://www.ons.gov.uk/census/censustransformationprogramme/questionde velopment/genderidentity/census2021finalguidanceforthequestionwhatisyoursex, last accessed 19/07/21.
35. Driver and Vehicle Licensing Agency response to freedom of information request, https://www.whatdotheyknow.com/request/117318/response/290540/attach/html/2/FOIR 2962%20Paul%20Wilson.pdf.html#:~:text=What%20are%20the%20basic%20colou rs%20stored%20in%20the%20DVLA%20vehicle%20database%3F&text=Beige%2C %20black%2C%20blue%2C%20bronze,%2C%20turquoise%2C%20white%20and %20yellow, last accessed 19/07/21.
36. B. Berlin and P. Kay, *Basic Color Terms: Their Universality and Evolution*, University of California Press, Berkeley, CA, 1969.
37. N. B. McNeill, Colour and colour terminology, *J. Linguist.* 8(1), 1972, 21–33. JSTOR, www.jstor.org/stable/4175133, last accessed 19/07/21

38. Debi Roberson, Jules Davidoff, Ian R. L. Davies and Laura R. Shapiro, Color categories: Confirmation of the relativity hypothesis. Web published paper 2005 http://research.gold.ac.uk/id/eprint/5673/1/PSY_davidoff-robertson-color-categories_2005.pdf, last accessed 19/07/21.

39. J. L. Hardy, C. M. Frederick, P. Kay and J. S. Werner, Color naming, lens aging, and grue: What the optics of the aging eye can teach us about color language, *Psychol Sci.* 16(4), 2005, 321–327. doi:10.1111/j.0956-7976.2005.01534.x

40. Bernard Comrie, Language and thought, Linguistic Society of America, https://www.linguisticsociety.org/resource/language-and-thought, last accessed 19/07/21.

41. Wikipedia, Lingusitic relativity, https://en.wikipedia.org/wiki/Linguistic_relativity, last accessed 19/07/21.

42. G. Deutscher, *Through the Language Glass: Why the World Looks Different in Other Languages*, Cornerstone, London, 2010.

43. Wikipedia, Proto-Sinaitic script, https://en.wikipedia.org/wiki/Proto-Sinaitic_script, last accessed 19/07/21.

44. *The Secret History of Writing*, BBC Four television broadcast, 2020. Presented by Lydia Wilson, written and directed by David Sington, https://www.bbc.co.uk/programmes/m000mtmj, last accessed 19/07/21.

45. Anshuman Pandey, Revisiting the encoding of Proto-Sinaitic in unicode, Web published paper 30/07/19, http://www.unicode.org/L2/L2019/19299-revisiting-proto-sinaitic.pdf, last accessed 19/07/21.

46. Y. N. Harari, D. Vandermeulen and D. Casanave, *Sapiens: A Graphic History - Volume 1: The Birth of Humankind*, Jonathan Cape, London, Penguin Random House, pp 86–96.

47. Radiopaedia, https://radiopaedia.org/encyclopaedia/signs/all, last accessed 19/07/21.

48. Agence France-Presse, Never say die? German bill using feminine word forms sparks row, *Guardian*, 13/10/20, https://www.theguardian.com/world/2020/oct/13/never-say-die-german-bill-using-feminine-word-forms-sparks-row, last accessed 19/07/21.

49. K. Rayner, S. J. White, R. L. Johnson and S. P. Liversedge, Raeding wrods with jubmled lettres: There is a cost, *Psychol. Sci.* 17(3), 2006, 192–193. doi:10.1111/j.1467-9280.2006.01684.x

50. R. Beale, Eeek!, *London Review of Books* 43(5), March 2021, 20–21.

51. The Swansea house that looks like Hitler, Wales Online, 29/03/11, https://www.walesonline.co.uk/news/wales-news/swansea-house-looks-like-hitler-1849045, last accessed 19/07/21.

52. C. Hadfield, *You Are Here: Around the World in 92 Minutes*, Pan Books, Basingstoke, 2015, p 48.

53. L. B. Alberti, *On Painting and on Sculpture (De Statua)*, C.Bartoli, Milan, 1568.

54. D. Hockney and M. Gayford, *A History of Pictures*, Thames & Hudson, London, 2016, p 34.

55. L. da Vinci, *A Treatise on Painting*. Translated by John Francis Rigaud, J.B Nichols and Son, London, 1835, https://archive.org/details/davincionpainting00leon/page/n5/mode/2up, last accessed 19/07/21.

56. Turing Alan Mathison, The chemical basis of morphogenesis, *Phil. Trans. R. Soc. Lond. B* 1952, 237(641), 37–72. doi:10.1098/rstb.1952.0012

57. Kat Arney, How the zebra got its stripes, with Alan Turing, *Mosaic Science*, 11/08/14, https://mosaicscience.com/story/how-zebra-got-its-stripes-alan-turing/, last accessed 19/07/21.

58. Ian Stewart, interview contribution to Alan Turing: Turing's final challenge, in The Secret Life of Chaos, *BBC Four* television broadcast, 2011, https://www.bbc.co.uk/programmes/b00pvlc3, last accessed 19/07/21.

59. P. Ball, Forging patterns and making waves from biology to geology: a commentary on Turing (1952) 'The chemical basis of morphogenesis', *Phil. Trans. R. Soc. B* 2015, 3702014021820140218. doi:10.1098/rstb.2014.0218

60. E. H. Gombrich, *Topics of Our Time: Twentieth-Century Issues in Learning and in Art*, Phaidon, London, 1991, p 180.

61. E. H. Gombrich, *Art and Illusion*. 5th edition, Phaidon, London, p 60.

62. Tom Goodwyn, Muse unveil artwork for new album 'The 2nd Law', *NME*, 31/07/12, https://www.nme.com/news/music/muse-188-1261239, last accessed 19/07/21.

63. *Muse Map Out the Brain on 'The 2nd Law' Album*, Rolling Stone, 30/07/12, https://www.rollingstone.com/music/music-news/muse-map-out-the-brain-on-the-2nd-law-album-art-192941/, last accessed 19/07/21.

64. M. Cobb, *The Idea of the Brain*, Profile Books, London, 2020.

65. R. Newman, *Neuropolis: A Brain Science Survival Guide*, William Collins, Glasgow, 2017.

66. Adam Forrest, Matt Hancock admits Hollywood film Contagion helped shape his vaccine response, *Independent*, 04/02/21, https://www.independent.co.uk/news/uk/politics/covid-vaccine-strategy-hancock-contagion-movie-b1796923.html, last accessed 19/07/21.

67. S. Abrahams, *Every Painter Paints Himself: Portraits of the Artist and An Alter Ego*. Blurb Books, London, 2010

68. Every Painter Paints Himself, https://www.everypainterpaintshimself.com/galleries/page/P45, last accessed 19/07/21.

69. A. Rothenberg, Rembrandt's creation of the pictorial metaphor of self, *Metaphor Symb.* 23(2), 2008, 108–129, doi:10.1080/10926480801944269

70. *Rembrandt Through His Own Eyes*, Oxford University Press, 15/07/10, https://blog.oup.com/2010/07/rembrandt/, last accessed 19/07/21.

71. D. E. McGlynn and R. B. Gunderman, Learning to see: the moral opportunity of art, *J Am Coll Radiol*. 11(6), 2014 Jun, 536–9. doi:10.1016/j.jacr.2013.06.011. Epub 2013 Aug 24. PMID: 23978735.

72. J. van Dijck, *The Transparent Body: A Cultural Analysis of Medical Imaging*, University of Washington Press, 2005, pp 51–53.

73. G. D. Schott, Exploring extramission: Plato's Intraocular "Fire" revisited, *Perception* 48(12), 2019, 1268–1270. doi:10.1177/0301006619884208

74. http://www.yorku.ca/rsheese2/1010/perception.htm

7 Pictorial Review

This chapter presents a selection of artworks either inspired by or strongly resonant with radiographic imagery or concepts. The intention here is to allow the works to – largely – speak for themselves accompanied by some brief contextualisation. While the works have been selected to reinforce themes explored elsewhere in the book and, as such, represent something of a cross-section of radiographic art, it is by no means intended to be a comprehensive survey. There are plenty of gems to admire. Exploration of the featured artists' back catalogue via an internet search is nevertheless strongly recommended.

The first works in the gallery do not, at first, appear to belong in this selection, lacking overt radiographic features or content. Let us consider *Ever After after Danloux and Van der Kooi* (2019) by Derrick Guild first of all, shown in Figure 7.1. The viewer is coaxed into synthesising the four locket-shaped oil paintings into one cohesive individual, but the oval paintings have been meticulously reproduced from not one but two old masters. On closer inspection, it becomes clear that while the eye may belong to the same face as the nose/mouth, the earring and clothing do not appear to be those of the same individual. Yet even once the spell is apparently broken, when our eyes return to the work it is difficult not to conjure up the individual we thought we saw to begin with. Guild plays with Gestalt type illusionism while probing the constraints and possibilities of viewing the action through a frame (albeit oval in this case rather than rectangular). We are gently reminded that in our collective tendency to construct fiction from the data in front of us, our perception of the world may veer towards fairy tale rather than reality.[1]

In the modified self-portrait *I* (illustrated in Figure 7.2) no amount of imaginative endeavour on the viewer's part can conjure a head for the subject. Alexandra Odaryuk drew on her experience of depression and anxiety to deliver this work, intended to raise awareness of mental health issues. The potentially hidden nature of depression is reflected in the invisibility of the head, but the gesture of the hands – raised in despair to an absent face – provides the viewer with some understanding of the anguish. By eliminating the most personal aspect of a portrait –- the face – Odaryuk encourages us to uncouple our assessment of an individual's inner state from superficial cues and signals. As a radiologist I spend my professional life looking at images of people's bodies in a fragmented, dismembered fashion – even a "whole body" MRI scan is typically acquired as multiple discrete components, each of which is composed of numerous individual slices. In the age of teleradiology the radiologist may even be on a different continent to the patient. While this is not necessarily to the detriment of the care delivered, works such as Odaryuks' remind me of the well-known clinical adage "don't treat the X-ray, treat the patient".

DOI: 10.1201/9780367855567-7

FIGURE 7.1   Ever After after Danloux and Van der Kooi (2019) by Derrick Guild. Courtesy of the artist and The Scottish Gallery.

Another work which at first seems far removed from radiology is Richard Mosse's *Incoming*, an ambitious and highly acclaimed response to the migrant crisis emerging in 2017, but of ongoing and escalating relevance at the time of writing. Mosse used a military-grade infrared camera to capture the movement of individuals undertaking perilous journeys from North Africa and the Middle East in search of safety in Europe. Two frames from the 52-minute installation are shown in Figures 7.3 and 7.4. The use of military hardware to generate images for artistic and humanitarian purposes was a self-conscious decision by Mosse, seeking to subvert the typical portrayal of migrants by governments.[2] Such "weaponised" images have strong resonance with the themes we explored in Chapter 5 – the viewer is forced to confront the possibility of complicity in the events unfolding, or at the very least to contemplate the seeming indifference of society. The use of image capture that relies on electromagnetic radiation outwith the visible spectrum is certainly akin to the activity undertaken in an

FIGURE 7.2   I by Alexandra Odaryuk. Courtesy of the artist.

X-ray department. Whilst the infrared camera does not reveal structures beneath the skin in the way X-ray-based techniques or MRI does, it shares with them the abstract conversion of flesh into greyscale images, in a fashion unrelated to reality as we usually perceive it. The images superficially resemble a negative image of a black and white photograph, but do not "normalise" when the greyscale is inverted.

This abstraction has two notable effects, both similar to the transformative image conversion encountered within radiology. The first is a neutralisation of skin tone, such that without additional cues the ethnicity of individuals cannot be determined. The removal of discriminating visual features may facilitate an evaluation of a humanitarian crisis uncoupled from any potential racial prejudice. In the two frames illustrated here, we see handprints left behind on the surface of dinghy, short-lived heat signatures captured by the infrared camera. The handprints resonate with those on cave walls we considered in Chapter 1, Bertha Röntgen's iconic left hand, and some additional immortalised hands we shall consider in the final chapter. Images of an individual's hand transcending time, geography, and ethnicity, reminding us of our common humanity.

The second effect is that in presenting the action in an unfamiliar fashion, we become distanced or detached from it – we are all the more aware that we are viewing a representation of reality rather than reality itself. Yet the detachment does not equate to disinterest. In some ways, this protective layer of detachment allows a greater level of scrutiny than might be achieved viewing the same images in conventional film or video. Likewise, within medical imaging the abstracted representations of internal anatomical reality, typically rendered in greyscale are perhaps more palatable to scrutinise on a daily basis than the reality of human flesh lacerated by trauma or ravaged by disease.

FIGURE 7.3   Richard Mosse Incoming (2017). © Richard Mosse. Courtesy of the artist and
Jack Shainman Gallery, New York.

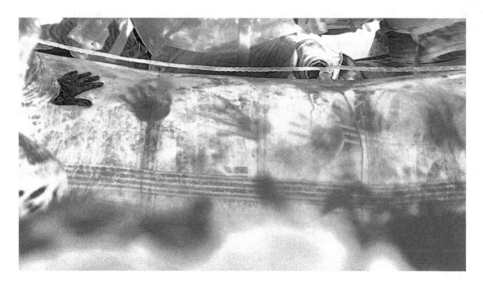

FIGURE 7.4   Richard Mosse Incoming (2017). © Richard Mosse. Courtesy of the artist and
Jack Shainman Gallery, New York.

Our next work is also a response to migration-related events. *Protest* (2010) by Marilène Oliver (shown in Figure 7.5) was, in her own words,

inspired by the story of an illegal immigrant who disembowelled himself in protest to his imminent deportation. The work is made from hundreds of layers of printed card. On the top side of each layer of card is a CT scan, on the underside is the UK Immigration act stained red, the intestines are shredded.[3]

FIGURE 7.5 Protest (2010) by Marilène Oliver. Courtesy of the artist

Most of Oliver's work draws very heavily upon medical imaging technology, using CT and MRI datasets as a template from which she shapes extraordinary sculptures. Contemplating the ever-increasing digitialisation of the body, her work has been exhibited widely in the UK and Europe.[4] Her early work utilised data from the Visible Human Project. A convicted murderer in the United States donated his corpse to medical science which, following his execution by lethal injection in 1993, was both scanned using CT and MRI and finely cryosectioned in the axial plane. The 1-mm thick slices were photographed and the images, together with the matching cross-sectional datasets uploaded to the internet as a public domain anatomical library.[5, 6] This project has echoes of earlier anatomical history when the bodies of executed criminals were the only authorised source of cadavers for dissection and rekindled some debate on the validity of consent in this setting.[7]

Oliver's more recent works, including *Protest*, have utilised an anonymised whole-body CT dataset of a young woman known as Melanix. Being of similar body shape to the artist, Oliver has used Melanix not only as her muse but "as a double, a stand in" for herself in her work.[3] This aspect of self-identification adds a further dimension of horror to *Protest* – we are compelled to recognise that the flesh being shredded could very well be our own under different circumstances, or equally horrifying, that an act of catastrophic self-harm is the result of laws enacted in our own name.

The political dimension of Oliver's CT-based artwork is therefore unequivocal, but utilises the body morphology of an individual unrelated to the tragic events that inspired it. By contrast, Ai Wei Wei's work *Brain Inflation* (2009)[8] could not be any more personal in how it conveys a politically charged message. In August 2009 the Chinese artist and activist was severely beaten by police in his hotel room in Chengdu, resulting in significant head trauma.[9] He had been intending to give testimony in the trial of writer Tan Zuoren, who had been investigating corruption contributing to the negligent construction of school buildings in the Sichaun region – thought to have been responsible for the deaths of some 5,000 children following earthquakes in 2008. In addition to the beating, Ai Wei Wei was detained for the duration of the trial and prevented from testifying (Tan was convicted in 2010 and spent four years in prison).[10] Shortly afterwards Ai developed neurological symptoms whilst in Munich, an MRI scan demonstrating a large subdural haematoma which required neurosurgical evacuation. *Brain Inflation* consists of unaltered images from this MRI examination, showing both the subdural haemorrhage and post-traumatic orbital soft tissue swelling.[9] The context here is all important – similar scan appearances could be produced by all sorts of accidental injuries, but awareness that this life-threatening trauma is the result of state-sponsored witness intimidation imbues the work with its power.

In displaying untampered MRIs Ai can be said to have delivered a "readymade" in the style of Duchamp – the Munich radiographers arguably being the original creators of the image. Our next work, whilst also an MRI-based self-portrait, required substantial stylistic handiwork on the part of the artist. *Brain of the Artist* (2012) by Angela Palmer (Figure 7.6) is composed of 16 sheets of glass placed in close proximity to one another in a cube-like arrangement. On each sheet Palmer has engraved the outline of her own brain structure as revealed by MRI, individual

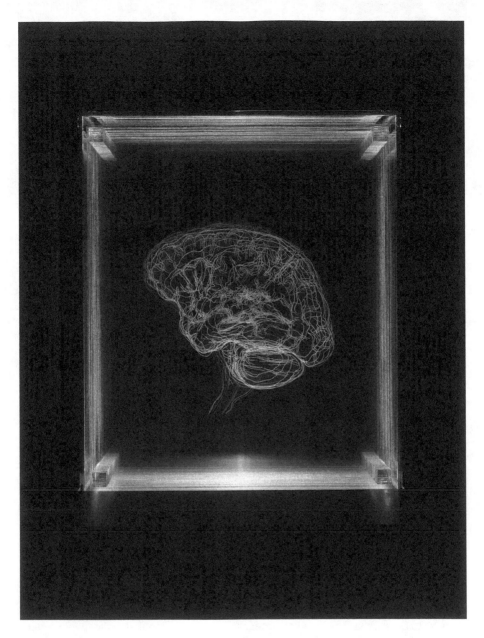

FIGURE 7.6   Brain of the Artist (2012) by Angela Palmer. Courtesy of the artist and National Galleries Scotland.

panes corresponding to sagittal scan slices.[11] As a three-dimensional glass sculpture, a printed illustration does not really do the work justice. Viewing it when exhibited at the Scottish National Portrait Gallery, the work had a holographic quality – from certain angles the brain takes on a tangible 3D appearance, whilst from other angles it almost disappears altogether. Palmer encourages us to consider both the beauty of objectively rendered anatomy and the importance of the viewer's vantage point. Produced in cuboidal glass, the work certainly resonates with the window theme discussed in Chapter 5.

In *X-Ray Architecture*, Beatriz Colomina draws attention to the prevalent use of glass in modern architecture and the related influence of radiographic concepts:

> Glass architecture echoes the logic of the X-ray. There is an outer screen that disappears in order to register a ghostly image of the inside. It is X-ray architecture. As with Röntgen's transformative images, X-ray architecture is an image of an image - the effect of an X-ray, rather than an actual X-ray. It's not so much that the inside of the building is exposed, but that the building represents exposure, and this exposure occurs on a screen. Glass is called on to simulate transparency.[12]

Two buildings designed by architects that Colomina suggests exemplify this trend are illustrated in Figures 7.7 and 7.8. The Glass House by Philip Johnson in New Canaan, Connecticut, (Figure 7.7) certainly lives up to its name – short of botanical glasshouses, it is hard to think of more transparent accommodation.[13] The Neue Nationalgalerie art gallery in Berlin, designed by Ludwig Mies van der Rohe (Figure 7.8) is likewise not a good place to go throwing stones. Completed in 1968, the roof of the gallery, composed of rows of repeating squares, also pre-empts the grid-based cross-sectional images that would follow in the 1970s.[14]

Ted Humble-Smith's work *1972* (shown in Figure 7.9) celebrates the arrival of the first cross-sectional modality, CT. This work was commissioned by the Royal Academy of Engineers as part of a series, *Fifty*, to celebrate a half-century of the prestigious MacRobert award, of which the CT scanner was an early recipient.[15] In this striking image, a seemingly rotating disc mimics the plane of the scanners acquisition plane, intersecting a semi-transparent human skull. As a dedicated Spielberg fan, this work reminds me of not one but two Indiana Jones movies – the intersecting disc resembling a rotating blade in one of the booby traps protecting the Holy Grail in *Indiana Jones and the Last Crusade* (dir. Steven Spielberg, 1989) and the skull not unlike the titular ones featured in *Indiana Jones and the Kingdom of the Crystal Skull* (2008). However, for viewers unencumbered by such cinematic clutter, the rotational motif works as a highly effective visual metaphor – not only alluding to the mechanism by which images are generated, but also the revolutionary impact CT had in medicine and surgery. The skull is positioned as it would be in a CT scanner, with the patient lying horizontally, which makes me want to rotate the image 90 degrees clockwise to see the skull "properly" – reminiscent of Holbein's anamorphic skull in *The Ambassadors*. As mentioned in Chapter 5, the acquisition plane for CT head scans is skewed by a few degrees from the "true" axial plane so the orbits can be excluded from the scan volume, another reminder of the plasticity of anatomical convention.

FIGURES 7.7 Photograph of The Glass House by Philip Johnson in New Canaan, Connecticut. By Ritu Manoj Jethani. Image credit: Shutterstock.

FIGURE 7.8 Photograph of The Neue Nationalgalerie art gallery in Berlin. By andersphoto. Image credit: Shutterstock.

The highly pixelated appearances of early CT and MRI scans share a visual heritage with mosaics and woven cloth as discussed in Chapter 1. The quantised, unitary image composition is also akin to – stay with me on this one – jigsaw puzzles. In normal circumstances, an important distinction is that the image is usually pre-existing before being arbitrarily fragmented. In view of the close connections between cartography and medical images (discussed in Chapter 4) it is noteworthy that the

FIGURE 7.9   1972 by Ted Humble-Smith. Courtesy of the artist and the Royal Academy of Engineers.

original name for jigsaws was "dissected maps". This term was coined by Berkshire map-maker John Spilsbury who in 1760 pioneered the first jigsaw puzzles in the form of maps of the British Empire cut into pieces.[16, 17] While at the time the use of "dissected" was probably a relatively neutral term for cutting things into pieces, to modern readers it is difficult not to associate it with anatomical dissection.

In her project *Within 15 minutes*, Alma Hasser separately photographed identical twins in the same outfits and poses, constructed (or maybe deconstructed) jigsaw puzzles from each portrait, and then swapped alternate pieces of each puzzle to produce two half-and-half hybrid portraits.[18] The result is fragmented, partially deconstructed approximations of individuals. An example is shown in Figure 7.10. Hasser's work explores the unique nature of human identity – monozygotic twins have historically been characterised as sharing an identical genome, yet differing environmental influences on twins start in utero, such that fingerprints vary between siblings. A recent study has suggested that while identical twins may start with the same DNA, there is a (very modest) divergence in genetic material during gestation such that at birth the genomes of identical twins vary by an average of five mutations.[19] Further genetic divergence will occur over the twins lifetime in the form of further mutations and the mechanisms of epigenetics, but this divergence is not unique to monozygotic twins. Every individual is subject to these processes, such that the genome we start with at the moment of conception is not the same by the end of our life. So the rebranded title of "A Human Genome Project" I suggested in

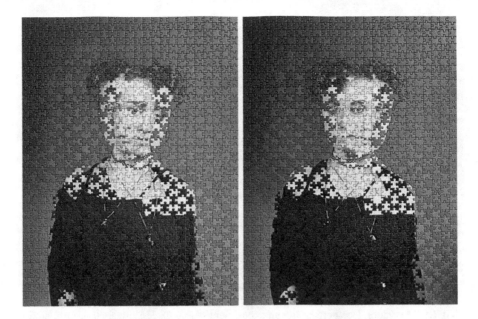

FIGURE 7.10   Within 15 minutes by Alma Hasser. Courtesy of the artist.

the previous chapter requires the additional caveat "at a particular moment in time" to reflect the ever-changing genetic landscape. It is a reflection of how far genetic sequencing technology has come in the past two decades that by comparison to the solitary genome completed as a worldwide effort in 2001 the paper on twins referenced above (one of the thousands published each year) involved the sequencing of an astonishing 49,962 individuals.[19]

In juxtaposing the two images side by side the viewer is invited to "spot the difference" like a puzzle from a child's magazine. As we saw in the previous chapter, radiologists are also in the business of spotting the difference, but as with Hasser's portraits, this is not always a straightforward undertaking. Even when the comparison is between images of the same individual, rather than a notional normal, there is usually extraneous noise to be filtered out, positional differences to factor in and technical differences between scanners to be compensated for.

Comparison between different versions is also required in the X-ray analysis of artworks. In recent decades the use of X-ray examinations to provide analytic information on pre-existing works of art has become a scientific discipline in its own right, including sophisticated, but high radiation dose techniques not suitable for use in humans. X-rays of masterpieces have revealed all sorts of new insights including compositional alterations, removal or additions of figures or furniture, and completely unrelated paintings (sometimes by different artists) which have been painted over altogether.[20] In the past few years works by Picasso,[21] Degas[22] and Da Vinci[23] have offered up previously unknown secrets through such analysis. Figures 7.11 and 7.12 provide an example. The painting, by 19th-century artist Henry Gillard Glindoni, depicts Queen Elizabeth visiting the house of polymath John Dee,

FIGURE 7.11   John Dee performing an experiment before Queen Elizabeth I. Oil painting by Henry Gillard Glindoni. Credit: Wellcome Collection. Attribution-NonCommercial 4.0 International (CC BY-NC 4.0).

FIGURE 7.12   John Dee performing an experiment before Queen Elizabeth I. Radiograph of oil painting by Henry Gillard Glindoni. Credit: Wellcome Collection. Attribution-NonCommercial 4.0 International (CC BY-NC 4.0).

accompanied by (amongst others) Sir Walter Raleigh on her left and Lord Treasurer William Cecil, just behind him. Dee is holding court demonstrating the explosive or flammable properties of some chemicals. Radiographic analysis performed in 2015 shows a circle of skulls surrounding Dee and his demonstration, subsequently painted over in the final composition.[24] Whenever previously hidden details such as these are revealed, the question arises whether the final, extant version of the artwork really is the definitive article – were the skulls an important part of Glindoni's original vision, perhaps mistakenly discarded at the behest of an interfering patron, or were they painted over with good reason?

Radiographic images of pre-existing artwork produce very striking images in their own right. Spanish artist Alejandro Guijarro has expanded on this in his 2016 series LEAD. Creating X-ray images of works by Van Dyck, Rubens, Delacroix, Goya, and Velazquez not for analytical purposes but as works of art in themselves.[25] One of the series, *73813 (Siege of Beauvais in 1472)* is reproduced in Figure 7.13, a radiograph of an original oil painting by François Louis Joseph Watteau. Reflecting on his work, Guijarro comments:

> At the heart of this series is a paradox: as x-rays they belong to the realm of scientific images, objective, possessing an unquestionable scientific truth. Yet, by their visual indeterminacy, they also exist in the subjective world of personal interpretation, the intuitional and emotional.[26]

FIGURE 7.13   73813 (Siege of Beauvais in 1472) by Alejandro Guijarro. This is a radiograph of an original oil painting by François Louis Joseph Watteau. Courtesy of the artist.

The paradox that Guijarro's work explores is by no means unique to X-ray images of artwork. Subjectivity, personal interpretation and indeterminacy are, as we have seen, equally applicable to clinical radiographs, or indeed any image for that matter.

Nevertheless, the transparency that radiographic images deliver does have both a powerful conceptual and aesthetic appeal. The work of "X-ray Artist" Hugh Turvey exemplifies both dimensions. While certainly not the first to perform radiographs for artistic purposes – flowers and other pleasing objects were X-rayed within months of Röntgen's discovery – Turvey has achieved international commercial and critical success for his body of work in which the elegance of familiar (and less familiar) objects and anatomy are uncovered through radiographic techniques. Permanent artist-in-residence at the British Institute for Radiology since 2009, Turvey has been actively engaged in educational activities, community-based projects, and hospital-based artwork designed to put patients at ease in radiology departments, in addition to work for global brands, such as Google, Ford, and Credit Suisse.[27, 28]

An example of his work is shown in Figure 7.14. The image is a radiograph (or "Xogram") of the costume featured in Zoe Philpott's one-woman show *Ada.Ada. Ada*. Celebrating the life and work of 19th-century computer pioneer Ada Lovelace, the costume features an LED array integrated into the show's lighting scheme.[29] The image demonstrates the luminous quality characteristic of Turvey's work, and has strong resonance with themes that have caught the public imagination from the earliest days of radiography. The power of X-rays to render fabric (as well as flesh) transparent stoked fears of voyeurism in the final years of the Victorian era, reflected in the marketing of lead-lined underwear. X-ray glasses subsequently proved popular in spy franchises from James Bond to Agent Cody Banks in the 1990s, and in 2011 modesty-preserving underwear made a return to protect travellers from the unwanted gaze of airport security scanners.[30] In the image of Ada's outfit no flesh or bodily structures are exposed, yet the image suggests the absent body highly effectively, radiographic contours providing a three-dimensional flavour absent in conventional photography.

X-ray images are often burdened by negative associations – a reminder of mortality perhaps, or cold, unwelcoming clinical efficiency. In this chapter, we have also seen examples of radiographic art delivering some devastating political and social critiques. In contrast, Turvey frequently employs exuberant, hand-touched colour schemes and includes humorous or witty touches to raise a smile. His work reminds us that while X-rays can certainly demonstrate disease, conflict, and self-mutilation, they can also provide inspiring images to lift the spirit – an unapologetic celebration of the beauty of radiographic images and all they can reveal.

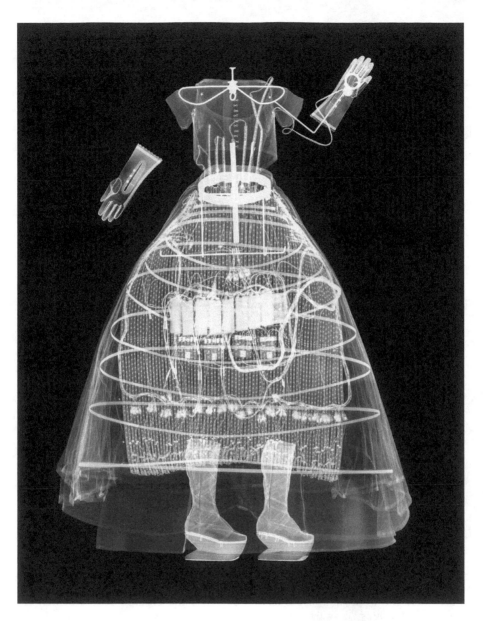

FIGURE 7.14   Ada.Ada.Ada by Hugh Turvey. The image is a radiograph (or "Xogram") of
the costume featured in Zoe Philpott's one-woman show Ada.Ada.Ada. Courtesy of the artist.

REFERENCES

1. *Ever After, Derrick Guild*, The Scottish Gallery, https://scottish-gallery.co.uk/exhibitio ns/ever-after, last accessed 20/07/21.
2. Incoming, Richard Mosse, http://www.richardmosse.com/projects/incoming#, last accessed 20/07/21.
3. M. Oliver, TED talk, quoted with permission from the artist, https://www.tedmed.com/talks/show?id=731053, last accessed 20/07/21.
4. Marilène Oliver, https://www.marileneoliver.com/, last accessed 20/07/21.
5. The National Library of Medicine's Visible Human Project, https://www.nlm.nih.gov/research/visible/visible_human.html, last accessed 20/07/21.
6. Wikipedia, Visible human project, https://en.wikipedia.org/wiki/Visible_Human_P roject, last accessed 20/07/21.
7. G. Roeggla, U. Landesmann and M. Roeggla, Ethics of executed person on Internet, [Letter], *Lancet*, 345(0), 28 January 1995, 260.
8. *Heather Sparks*, Science Sparks Art, 28/05/2014, https://sciencesparksart.tumblr. com/post/87140235698/brain-inflation-2009-ai-weiwei-this-mri-scan, last accessed 20/07/21.
9. Kristen Allen, Chinese artist gets emergency brain surgery in Munich, *The Local*, 16/09/21, https://www.thelocal.de/20090916/21969, last accessed 20/07/21.
10. Wikipedia, Tan Zuoren, https://en.wikipedia.org/wiki/Tan_Zuoren, last accessed 20/07/21.
11. National Galleries Scotland, https://www.nationalgalleries.org/art-and-artists/129231/ brain-artist, last accessed 20/07/21.
12. B. Colomina, *X-Ray Architecture*, Lars Müller Publisher, Zurich, 2019, p 135.
13. Ibid, p 142.
14. Wikipedia, Neue_Nationalgalerie, https://en.wikipedia.org/wiki/Neue_Nationalga lerie, last accessed 20/07/21.
15. Ted Humble-Smith, https://humblesmith.net/fifty, last accessed 20/07/21.
16. William Hartston, Top 10 facts about jigsaw puzzles, *Express*, 02/11/15, https://www. express.co.uk/life-style/top10facts/616404/Top-10-facts-jigsaws-puzzles#, last accessed 20/07/21.
17. Stewart T. Coffin, The puzzling world of polyhedral dissections, *John Rausch*, https:// johnrausch.com/PuzzlingWorld/chap01.htm#:~:text=Jigsaw%20Puzzles,is%20the%2 0familiar%20jigsaw%20puzzle, last accessed 20/07/21.
18. Alma Haser, http://www.haser.org/within-15-minutes, last accessed 20/07/21.
19. H. Jonsson, E. Magnusdottir, H. P. Eggertsson et al. Differences between germline genomes of monozygotic twins. *Nat Genet*. 53, 2021, 27–34. doi:10.1038/s41588-020-00755-1
20. BBC News, Nativity scene found under painting using X-ray, *BBC News*, 19/12/19, https://www.bbc.co.uk/news/uk-england-50844000, last accessed 20/07/21.
21. Nicola Davis, Artwork hidden under Picasso painting revealed by x-ray, *Guardian*, 17/02/18, https://www.theguardian.com/science/2018/feb/17/artwork-hidden-under-p icasso-painting-revealed-by-x-ray?, last accessed 20/07/21.
22. Ian Sample, X-ray reveals mysterious face hidden beneath Degas' portrait of a woman, *Guardian*, 04/08/16, https://www.theguardian.com/artanddesign/2016/aug/04/x-ray-reveals-mysterious-face-hidden-beneath-degas-portrait-of-a-woman?, last accessed 20/07/21.
23. Neil Smith, Leonardo da Vinci's abandoned and hidden artwork reveals its secrets, *BBC News*, 15/08/19, https://www.bbc.co.uk/news/entertainment-arts-49356073, last accessed 20/07/21.

24. John Dee performing an experiment before Queen Elizabeth I. Oil painting by Henry Gillard Glindoni, Wellcome Collection, https://wellcomecollection.org/works/nydjbrr7, last accessed 20/07/21.

25. Alejandro Guijarro, http://www.alejandroguijarro.com/works, last accessed 20/07/21.

26. Tristan Hoare Gallery, https://tristanhoaregallery.co.uk/artists/31-alejandro-guijarro/biography/, last accessed 20/07/21.

27. Hugh Turvey, https://www.x-rayartist.com/, last accessed 20/07/21.

28. British Institute of Radiology, https://www.bir.org.uk/about-us/artist-in-residence.aspx, last accessed 20/07/21.

29. Ada.Ada.Ada, https://adatheshow.com/, last accessed 20/07/21.

30. Yakub Qureshi, Inventor creates X-ray proof underpants to protect modesty of passengers in 'naked' airport scanners, *Manchester Evening News*, 15/03/11, https://www.manchestereveningnews.co.uk/news/greater-manchester-news/inventor-creates-x-ray-proof-underpants-856242, last accessed 20/07/21.

8 Imaging, Immortality, and Imagination

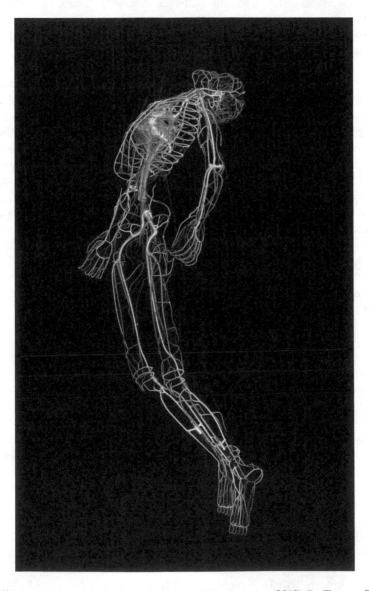

What will be remembered in the face of all that will be forgotten (2018). By Tavares Strachan. Courtesy of the artist.

DOI: 10.1201/9780367855567-8

In this final chapter we will contemplate mortality and immortality, the transient and the eternal. We will also consider the nature of information, the purpose of extracting meaning from data, and speculate on the future of medical imaging.

MORTALITY

From the very first image of a human hand, radiographic images and techniques have been strongly associated with human frailty and mortality. "I have seen my death" is said to have been Anna Bertha Röntgen's reaction on seeing the now iconic image of her skeletal hand.[1] Wilhelm Röntgen, in turn, was alarmed that the seemingly ghostly images and associations that radiographs conjured in the general population – at a time when spiritualism was enjoying increasing popularity – would undermine his scientific work.[2] Radiographic-style skeletons certainly took hold as a ubiquitous visual motif, and the arrival of cinema in the same year as X-rays were discovered helped to consolidate and disperse such imagery within the popular imagination, including *The X-ray Friend* (1897) and *The X-ray Mirror* (1899).[3] Another early example is available to view online via the BFI archive.[4] George Albert Smith was a British film pioneer based in Hove, who directed a one-minute silent film known as *X-rays* in 1897. The scene shows a couple sitting on a bench, the man attempting some flirtatious advances, welcomed by the woman at the start. A man carrying a movie-style camera (with "X RAYS" written on the side) emerges from the right side of the screen at which point Smith uses a jump cut to show the couple as skeletons (with the cloth of the woman's parasol also rendered invisible, leaving only the metal frame). When the cameraman replaces the lens cap a second jump cut transforms the couple (and the parasol) back to their normal appearance. The cameraman exits, but by this time the woman has grown tired of the amorous advances, gives the man a slap and also departs. While channelling the public fascination with newly discovered X-rays, the skeletal imagery utilises long-established associations of mortality from *Vanitas* and *Memento Mori* artworks. The couple's antics are pretty tame by modern standards but the film can be considered an early example of cinema coupling sex and death, taken to more graphic extremes in later horror movies. The voyeuristic potential (imagined or otherwise) of X-rays was certainly not lost on Victorian society, as mentioned in the previous chapter.

The innovation which puts Smith's skeletons into a distinctly radiographic frame is the instantaneous nature of the transformation – one moment we see the couple as we normally would, the next – without any intermediate transition – we see their skeletons (the reversal to normality also being instantaneous). This rapid to-and-fro transition is utilised in the near-ubiquitous electric shock motif, beloved of cartoons and CGI animated movies. Table 8.1 lists some examples.

This list is by no means exhaustive (and the quality of titles included is somewhat variable), but it certainly demonstrates the widespread use of the motif in popular family movies. Casting your eye down the list of directors (Steven Spielberg, George Lucas, Brad Bird, Robert Zemeckis, Chris Columbus, Wes Anderson) reveals this trope to have been employed by some of the most commercially successful directors of all time.

TABLE 8.1

Examples of the Radiographic Electric Shock Motif in Hollywood Movies

Film	Year	Director	Character/victim	Means of electrocution
Ghostbusters	1984	Ivan Reitman	Dana Barrett, Louis Tully	Demonic lightning strikes
Who Framed Roger Rabbit?	1988	Robert Zemeckis	Roger Rabbit	Fingers in electric socket
The Little Mermaid	1989	Ron Clements, John Musker	Ursula	Open to debate
Hot Shots!	1991	Jim Abrahams	Topper Harley	Loose electrical wiring
Home Alone 2: Lost in New York	1992	Chris Columbus	Marv Merchants	Electric burglar trap
Star Wars: Episode I	1999	George Lucas	Droid soldier	Gungan electroball in battle scene
The Stitch Movie	2003	Roberts Gannaway, Tony Craig	Stitch	Lightning strike
Ratatouille	2007	Brad Bird	Remy the rat	Lightning strike
Cloudy with a chance of Meatballs	2009	Chris Miller, Phil Lord	Flint Lockwood	Power transformer station misadventure
Fantastic Mr Fox	2009	Wes Anderson	Mr Fox, Kylie	Electric fence
Monsters vs Aliens	2009	Conrad Vernon, Rob Letterman	Susan Murphy	Experimental body shrinker
Top Cat: The Movie	2011	Alberto Mar	Benjamin Ball	Lightning strike
The Adventures of Tintin	2011	Steven Spielberg	Captain Haddock	Lightning strike
Guardians of the Galaxy	2014	James Gunn	Gamora	Electric ray in battle scene
The Boxtrolls	2014	Graham Annable, Anthony Stacchi	Boxtrolls	Fork in toaster
The Nut Job	2014	Peter Lepeniotis	Surly	Loose electrical wiring
The Incredibles 2	2018	Brad Bird	Electro	Super power self-electrocution
Tom and Jerry	2021	Tim Story	Tom	Sabotaged electrical cable

The radiographic electric shock has also, of course, has made prolific appearances in other cartoons, TV shows and elsewhere. A recent, rather bizarre example was cartoon Happy Meal characters in a TV advert for McDonald's. While the visual connection of X-rays and mortality is potentially unfortunate, it is worth pointing out that survival is the rule in these cartoon electrocutions, so perhaps the association is more positive than might be supposed. The motif certainly exemplifies the interchange of an idea through different visual media and disciplines, the radiographic transferred to the cinematograph, then to cartoons and back to the cinema screen in both cartoon and CGI animation.

Computer-generated animation is also more deeply enmeshed within medical imaging than simply rehashing the radiographic electric shock. In 1972 Ed Catmull and Fred Parke produced the first computer-generated animation of a human hand. Following in the tradition of cave stencils and Bertha Röntgen's radiograph, Catmull used his left hand as the basis for the images, labelling a plaster cast with 350 polygons which were then laboriously registered onto a computer program to produce a striking one-minute film in which the inert plaster cast comes to life with a full range of digital movements.[5] The animation was featured in the 1976 movie *Futureworld* (dir. Richard T. Heffron) and can also be viewed online.[6] Catmull would become a co-founder of PIXAR and eventually president of Disney. In addition to the groundbreaking computer-aided animation techniques that would lead to the first-ever full-length CGI feature film, *Toy Story* (dir. John Lasseter, 1995), PIXAR also pioneered the software algorithms for producing 3D reconstructions of CT scans.[7] Figure 8.1 shows a CT image of a hand, generated using this type of software. It bears a very strong resemblance to the hand of lead character Miguel in a scene from the Disney movie *Coco* (dir. Lee Unkrich, 2017). The flesh of Miguel's hand becomes largely transparent as his body occupies an uncertain intermediate state between the realms of the living and the dead. There is a similar blurring of the boundaries between creative and scientific endeavour in such images, reliant not only upon robust physics, mathematics and engineering, but also artistic talent and aesthetic choices.

Returning to our cartoon electric-shock skeletons, they are frequently depicted in both a white bone/black background and reversed into black bones on a white background, resonating with the image inversion we grappled with in Chapter 1. In addition, the white bones typically have a slightly blue colouring around the edges, rather than pure white, reflecting both the blue seen in electric sparks and lightning strikes, but also reminiscent of a subtle blue tinge present in many hard-copy radiographs as traditionally viewed using a viewing box. The use of the motif in the recent *Tom and Jerry* movie, reflecting countless uses in the original cartoons, utilises both the black/white inversion and blue tinge features. In the realm of popular culture, it is notable that hard-copy radiographs have endured beyond the length of their clinical shelf-life; traditional X-ray films and viewing boxes are now a rare sight in Western hospitals but are still commonplace in film and TV programmes (likely to relate to the frequency band artefact seen on computer monitors when filmed). In the early days of radiography natural light was harnessed to view X-ray films. Figure 8.2 shows a device used to reflect daylight onto a viewing frame in which the radiograph was secured. To modern eyes, it strongly resembles a laptop or tablet device. An alternative scheme for viewing radiographs suggested in the late 19th century is shown in Figure 8.3. Here the images are secured directly against a window frame, natural daylight again supplying the illumination. In my working life, I am typically starved of natural daylight, spending many hours in windowless rooms, so this looks

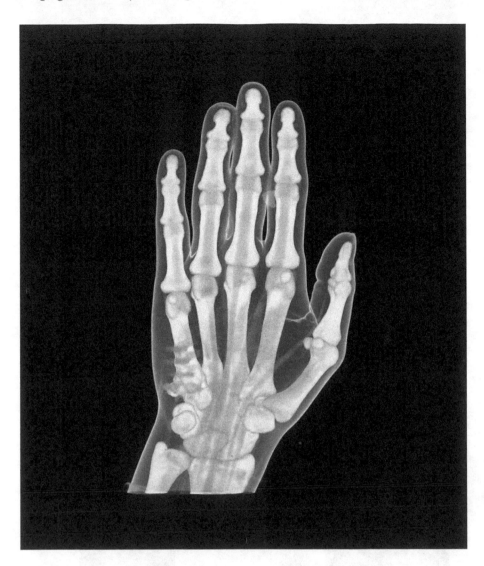

FIGURE 8.1   3D CT reconstruction of a hand. Image provided by the author.

like quite an appealing arrangement, albeit not a practical one in the era of thousands of images to look at. The arrangement also has a pseudo-religious dimension, resembling a stained-glass window in a church. I have juxtaposed a photograph of Eduardo Paolozzi's Millenium Window (2002) in St Mary's Episcopal Cathedral in Edinburgh (Figure 8.4), the last major work of the pioneering Scottish artist before his death in 2005.

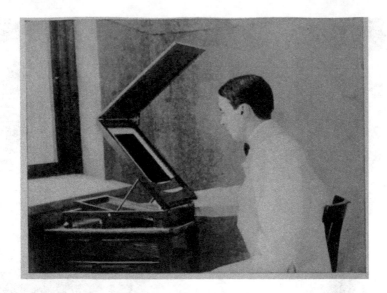

FIGURE 8.2   Illustration of a device used to reflect daylight onto a viewing frame in which the radiograph was secured, 1902. Credit: The Röntgen rays in medicine and surgery as an aid in diagnosis and as a therapeutic agent designed for the use of practitioners and students/ by Francis H. Williams. Wellcome Collection. Public Domain Mark.

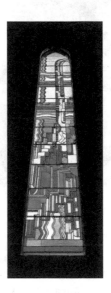

FIGURE 8.3   Radiographs secured directly against a window frame, natural daylight again supplying the illumination. Credit: Die Röntgentechnik: Lehrbuch für Ärzte und Studierende/von Dr. Albers-Schönberg, 1906. Wellcome Collection. Public Domain Mark.

FIGURE 8.4   A photograph of Eduardo Paolozzi's Millenium Window (2002) in St Mary's Episcopal Cathedral in Edinburgh. Image provided by the author.

IMMORTALITY

Images, whether created by great artists, medical examinations, or simply holiday snaps have the capacity to outlive both their creators and their subjects, and in doing so achieve a form of immortality. Doctors, so the rather grim aphorism tells us, bury their mistakes. For radiologists working in the modern era, mistakes are not buried but immortalised in the digital archive, preserved indefinitely for future scrutiny. Indeed, physicists inform us that information can never be destroyed, meaning all mistakes (professional or otherwise) are out there for eternity, long after the digital archive has ceased to exist, a disconcerting prospect for most of us.[8] Some of the artworks we considered in the previous chapter explored the recent collision of physical bodies with digital information technology. Human bodies have, however, achieved forms of immortality long before the advent of digital archives.

Figure 8.5 shows the body of a victim of the volcanic eruption that consumed the city of Pompeii in 79 CE. However, it is not the body itself, but a cast produced by a technique developed by 19th-century archaeologist Giuseppe Fiorelli in which liquid plaster was poured into the hollows left in the ash after the body had decomposed.[9] The cast occupies the cavity previously occupied by the victim's body – empty space inverted into a solid object. Information provided by the morphology and dimensions of the cavity provides us with a moving testament of this human tragedy, despite the physical absence of a body.

FIGURE 8.5   Photograph of a victim of the Mount Vesuvius eruption in Pompeii, preserved as a plaster cast. Image credit: Shutterstock

As a medical student, I remember reading a passage in the *Oxford Handbook of Clinical Specialties,* a near compulsory textbook carried about on the wards at the time.[10] It described human images being burnt into stone during the bombings of Hiroshima and Nagasaki, emphasising the enduring nature of the idea of these bodies rather than any specific (all too transient) human body. It struck me at the time to be an extraordinarily bleak passage to send fresh-faced medical students onto the wards with, but the concept is a powerful one, and has acquired additional resonance since embarking on a career in radiology.

In the mesmerising documentary *Can't Get You Out of My Head* (2021), Adam Curtis highlights the tragic demise of cosmonaut Vladimir Komarov, who in 1967 embarked on a spaceflight to mark the 50th anniversary of the Soviet Union. Despite being all too aware of some 203 faults in the spacecraft listed by his friend Yuri Gargarin, Komarov knew that Soviet high command would never countenance cancellation of the flight and that Gargarin would be forced to participate if he stood down. Resigned to his fate, Komarov climbed aboard the *Soyuz 1* but insisted that an open casket was used at his funeral in the event of his death. His incinerated remains, more closely resembling volcanic rock than a human body, formed a sobering indictment of intransigent political expediency.[11]

Let us briefly detour from immortalised human bodies to those of dinosaurs. At the climax to the movie *Jurassic Park III* (dir. Joe Johnston, 2001) paleontologist Alan Grant (played by Sam Neill, reprising this role from the original movie) manages to save himself from being eaten by a pack of hungry-looking velociraptors in an inventive, if eye-brow raising fashion. Earlier in the film, one of his colleagues uses a 3D printer to produce a plastic model of a velociraptor voice box, using the imaging data from a CT scan of a velociraptor fossil. Grant fishes the model out of his pocket just as the dinosaurs are about to attack and manages to distract them by blowing into it, producing some sort of velociraptor-like sound. I remember thinking this was highly far-fetched when I saw the film at the cinema (the bringing dinosaurs back to life bit was obviously fine!). However, in a life-imitating-art development, a broadly similar endeavour (alright, no live dinosaurs involved this time) was reported in 2015. Digital performance artist Courtney Brown initiated a project in which CT data from the fossilised skull of a *Corythosaurus casuarius* (a duck-billed dinosaur from the late Cretaceous era) was used to 3D print the larynx and nasal passages, which was then used to recreate the noise the creature made in life (or some approximation) by blowing air through the model.[12]

A similar undertaking was reported by archaeological researchers in 2020, this time using an Egyptian mummy. Researchers from Royal Holloway, University of London, the University of York and Leeds University used the CT scanner at Leeds General infirmary to scan a 3,000-year-old mummified Egyptian priest, Nesyamun. Creating a model of the larynx using a 3D printer, they synthesised a vowel sound intended to be similar to the voice of Nesyamun.[13] The sound was reported to sound more like a sheep's bleat, but, undaunted, the team intend to recreate full sentences in Nesyamun's voice using computer models.[14] My hunch is that the anatomy of a 3,000-year-old Egyptian will not be all that different to Egyptians alive today, and given the numerous additional factors that determine how a voice sounds such as local accents, I'm not convinced this is a good use of research time or funding. To paraphrase another *Jurassic Park* character, they were too busy thinking about whether they *could*, they didn't stop to think whether they *should*.

While I may have some reservations, this project does provide a striking example of the enduring nature of anatomical information. Should a reconstruction of a human larynx be required from an early 21st-century subject in the future, no physical remains will be required. So long as a dataset of a cross-sectional imaging examination of the neck (CT or MRI) is still available (together with a functioning computer and 3D printer) no contact with the subject, living or dead, would be required. The digital archive can bestow anatomical immortality, albeit of a form unlikely to be of much use to the subject concerned.

ROPES, RINGS, AND THINGS

In Chapter 6 we considered some *écorché* figures based on the iconography of St Bartholomew in which the flayed skin takes on an appearance closer to that of a cloak, or sheet of cloth. Vesalius' title *On the Fabric of the Human Body* (*De humani corporis fabrica*) seems very appropriate. Now consider four more anatomical figures shown in Figures 8.6, 8.7, 8.8, and 8.9. These are arranged in chronological order, each showing a male *écorché* figure holding a length of rope – another recurring anatomical motif. Have a closer look at Figure 8.8 in particular – I'm wrong to describe this figure holding a length of rope, it is actually another example of flayed skin and perhaps belongs back in Chapter 6. Looking at the segment of skin between the figure's hands, however, it is at best twisted (as though wringing out a towel), and at worst knotted as though the skin may be used for some unspecified rope-like purpose. There is certainly some ambiguity determining what is in the realm of the body and what is artefactual. This ambiguity also reflects seemingly contradictory views of the body within the history of Christian culture, being both a corruptible, transient vessel for the immortal soul ("for dust thou art, and unto dust shalt thou return"), yet also as God's ultimate act of creation – an entity deserving of reverence, needing to be preserved intact for the day of resurrection (accounting for the abhorrence of anatomical dissection and robust efforts to prevent body snatching).[15]

While not grappling with such spiritual dimensions, similar ambiguities of what belongs to the body and what belongs to the world at large are at play when artefacts from an individual's culture or environment make an appearance in radiological images. The very first radiograph shows Bertha Röntgen's wedding band – a striking reminder of her life and individuality (or alternatively male-dominated society and ownership). In the early years of radiography hand radiographs became a novelty alternative to a portrait, and the fashionable women of

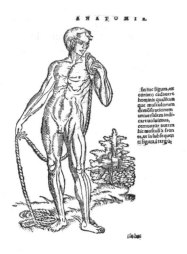

FIGURES 8.6 Male anatomical figure, holding a length of rope. Credit: Front of male figure, showing muscles. Wellcome Collection. Attribution 4.0 International (CC BY 4.0).

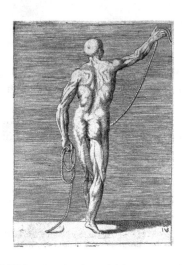

FIGURE 8.7 An *écorché* figure seen from the back, holding a length of rope. Engraving by G. Bonasone. Credit: Wellcome Collection. Attribution 4.0 International (CC BY 4.0).

FIGURE 8.8 An *écorché* figure seen from the back, holding something resembling a length of rope or fabric, but most likely representing the figure's flayed skin. Credit: Myotomia reformata: or an anatomical treatise on the muscles of the human body [William Cowper]. Wellcome Collection. Attribution 4.0 International (CC BY 4.0).

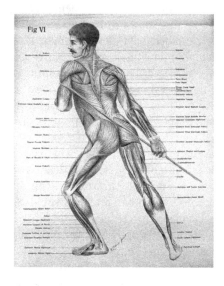

FIGURE 8.9 Artistic poses: a man, *écorché*, pulling a rope. Image credit: Lithograph by Robert J. Colenso. Wellcome Collection. Attribution 4.0 International (CC BY 4.0).

New York are said to have intentionally adorned numerous rings and bracelets as a style statement.[16] By contrast, for clinical examinations radiographers endeavour to remove jewellery – aspiring to an objective or neutral image devoid of life's clutter (or at least on a practical level preventing stuff getting in the way of seeing important structures). Yet the subjects of these examinations are not sterile laboratory specimens but individuals with a life beyond the examination room, and signs of individuality often manage to find a way into the images. In contrast to jewellery, foreign bodies such as embedded bullets and shrapnel indicate less fortunate encounters with the external environment. We have already seen that young children are prone to swallowing coins, and all sorts of other small radio-opaque foreign bodies have been captured on radiographs – either stuck within or passing through the gastrointestinal tract. In a homage to the "wound man" illustration we encountered in Chapter 5 (Figure 5.23), I have produced a composite image showing a selection of items (Figure 8.10). These items may not belong to the body (or belong within it for that matter) but when captured in this fashion they certainly constitute a part of the body's story.

An example of foreign body ingestion *par excellence* is shown in Figure 8.11. The efforts of artists such as Cezanne and Duchamps to capture motion or the 4th dimension of time within a two-dimensional image is perhaps trumped by this single abdominal radiograph in which the passage of time is indelibly written into the image. A young child ingested an impressive 24 individual toy magnets, shown snaking their way through successive bowel loops like a set of articulated train carriages. The position and configuration of the magnets indicates the magnets have progressed out of the stomach into the small bowel loops but have not got as far as the colon. The image transcends the 4th dimension, being an instant in time at the point the radiograph was acquired, yet revealing the past events of a magnet munching session and predicting the future of the child, requiring skilful surgical removal of the magnets (unlike most ingested foreign bodies which pass without complication, magnets can damage or perforate the bowel wall). Happily the child made a full recovery, so while the cautionary aspects of the image are self-evident, the associations many viewers will have of cartoon-type imagery of children or other creatures ingesting inappropriate meals can be enjoyed guilt-free. I have arranged Duchamp's *Nude Descending a Staircase, No.2* (1912) (Figure 8.12) and *Photographs of a Horse in Motion* by Eadweard Muybridge (1877) (Figure 8.13) alongside this radiograph as similar examples of dynamic activity captured in static images.

"Real-time" dynamic assessment is widely utilised in medical imaging from video-swallows in fluoroscopy, observation of bowel movement in ultrasound and MRI small bowel studies, angiographic assessment using DSA, CT techniques and Doppler ultrasound, and movement of the heart in echocardiography and "4D" CT and MRI studies. There is, nevertheless, something very powerful about the ever-moving, ever-changing body frozen in time in a single image. This is certainly the case in the striking images of Elizabeth Heyert's photography project *The Sleepers*, an example of which is illustrated in Figure 8.14. Heyert captured her subjects asleep

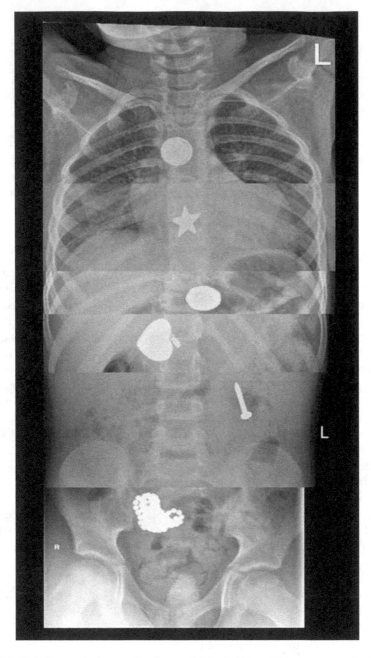

FIGURE 8.10   Composite image showing a selection of ingested items captured on chest or abdominal radiographs. Image provided by the author.

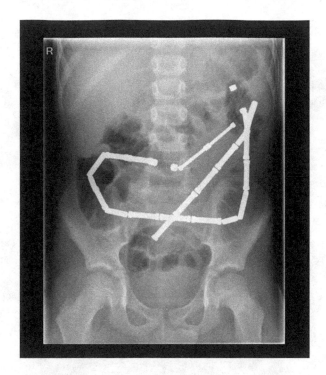

FIGURE 8.11 Abdominal radiograph of a young child showing 24 ingested toy magnets, shown snaking their way through successive bowel loops. Image provided by the author.

FIGURE 8.12 Nude Descending a Staircase, No. 2 (1912) by Marcel Duchamp. Image credit: Shutterstock

FIGURE 8.13 Photographs of a Horse in Motion by Eadweard Muybridge (1877). Image credit: Shutterstock

FIGURE 8.14   THE SLEEPERS, #15 photograph by Elizabeth Heyert, 2019. Courtesy of
the artist.

and unclothed, both at peace yet also in a moment of extreme vulnerability. There is some similarity to the casts of the Pompeii victims, perhaps accentuated by the stone-like appearance of the flesh. This transforms these all-too vulnerable figures into timeless statues.

The resemblance of these figures to ancient Romans reminds us that, as with Egyptian larynxes, human body morphology has not altered significantly during recorded human history. Some artefacts captured on radiographic images to this day, including coins and items of jewellery, would be recognisable to our ancestors, whether Egyptian, Roman, or Victorian. Perhaps even the pair of scissors embedded within a child's skull following an unfortunate accident shown as a CT 3D reconstruction (Figure 8.15) might be identifiable. Again, this child made a full recovery following neurosurgical removal of the scissors. Other artefacts are more era specific. Button batteries were not in existence in 1895 and I hope will not remain in circulation for much longer in view of the horrific damage they can cause to young children when lodged in the oesophagus following accidental ingestion. Specific types of ingested coins may pinpoint the date of a radiograph to a particular decade or two, and as digital currency replaces physical cash the presence of any coin on a radiograph may, in the future, narrow the period of acquisition to a relatively brief era of 150 years or so. Occasionally, artefacts on an imaging investigation may be able to pinpoint a patient to a specific date and location. Take a look at Figure 8.16 showing another 3D reconstruction of a CT scan, this time of a patient's chest. We can see the patient's lungs but their clothing is also visible with embroidered lettering faintly seen, marking a charity football match between Liverpool "Legends" against "Milan Legends".

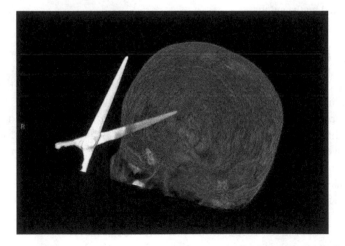

FIGURE 8.15 3D reconstruction of a CT head showing one blade of a pair of scissors embedded within a child's skull. The image is highly abstract with the portion of the blade embedded within the skull visible through a semi-transparent rendering of the skull bones. Image provided by the author.

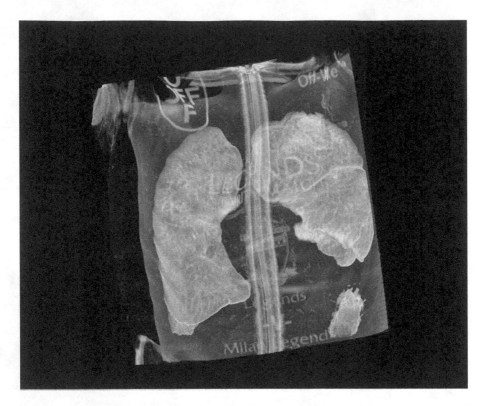

FIGURE 8.16 3D reconstruction of a CT chest scan. The lungs are visible but clothing is can also be seen with embroidered lettering including the Liverpool Football Club logo. Image provided by the author.

WAYS OF VIEWING

The culture of the day therefore manifests itself in radiographic images, but likewise cultural influences may impose particular interpretations or associations upon those images. Our understanding of the internal structure of the body is also intimately related to the specific methods by which we view it, which have transformed dramatically over the ages. Revealing the interior of the body has long been recognised as a source of spectacle, reflected in the construction of steep-raked viewing arenas in the renaissance era, such as the magnificent Anatomical Theatre of Padua, inaugurated in 1595, and longest surviving example of such a structure. A model of this is illustrated in Figure 8.17. The educational opportunities offered by opening up live bodies during operative procedures likewise turned surgery into a spectator sport, with similar steep-raked venues persisting well into the 20th century, such as that shown in Figure 8.18. While in North America, patients requiring surgery in the modern era are taken to the Operating Room (or "O.R.") in the UK we still talk of taking patients to "theatre".

FIGURE 8.17 Model of the Anatomical Theatre of Padua. Credit: Anatomical theatre at Padua, Diorama. Wellcome Collection. Attribution 4.0 International (CC BY 4.0).

FIGURE 8.18 Photograph showing (almost exclusively male) members of British Orthopaedic Association watching an operation performed by Vittorio Putti and staff at the Rizzoli Institute, Bologna,1924. Credit: Wellcome Collection. Attribution 4.0 International (CC BY 4.0).

The periscope arrangement shown in Figure 8.19 is (as far as I can tell) a suggestion or prediction of a potential educational device rather than something that was commonly used, but the projection of the interior of the body onto a screen visible not only by the surgeon but also the rest of the theatre team and trainees prefigures the development of laparoscopic surgery, which would facilitate a similar arrangement. In 2015, orthopaedic surgeon Timothy Miller broadcast an Achilles' tendon repair operation live for public viewing on the internet, one of many similar online surgical broadcasts made in recent years.[17] The name of the video-sharing application Miller used was, appropriately enough, *Periscope*. Figure 8.20 is a cartoon from 1932 predicting remote healthcare consultations conducted via TV screen, albeit in a tongue (not quite) in cheek fashion. While such tele-consultations have, in one form or another, occurred for several decades now, the pandemic of 2020 brought screen-to-screen online consultations into the mainstream.[18]

Figures 8.21 and 8.22 show two contrasting illustrations of the radiological gaze. In Figure 8.21, an illustration from an early radiology textbook, look how close the (exclusively male) doctors are huddled around their patient – the use of X-ray screening as depicted here is almost a direct continuation of the physical examination. In the more contemporary scene shown in Figure 8.22, our radiographers (both female) are separated from the patient by a lead-treated window as a protective safety measure and while they can see the patient directly (or their feet at any rate), the scan itself is viewed on computer monitors.

In recent years, the use of virtual reality technology alongside augmented reality and new methods of displaying cross-sectional imaging such as holograms is facilitating ever more immersive and detailed representations of the body. These can facilitate complex procedures, such as the use of VR by surgeons at Great Ormond Street Hospital to assist visualisation of blood vessels and brain structures when planning the surgical separation of conjoined twin girls whose skulls were fused. 3D printed models of the skull were also constructed to allow practice runs of this two-stage procedure successfully performed in 2018 and 2019.[19] Digitised anatomical information are now also used to create spectacular encounters with the body's inner workings, such as *The Tides Within Us* exhibition, touring Europe in 2021. Data from functional MRI scans are used to demonstrate the passage of oxygen through the body with extraordinarily dynamism and colour, with multiple screens displaying tissues and organs scaled as though the body were 600 ft tall.[20] A photograph of the exhibition is shown in Figure 8.23. Resembling both a fireworks display and a rapidly shifting tour through distant galaxies, this representation of both anatomy and physiology is at once unique and pioneering, whilst also very much continuing the tradition of the human body as spectacle.

Numerous innovative techniques for viewing the body have been predicted, pioneered, or promoted within the creative arena prior to being adopted within surgical practice or medical imaging. Before *Jurassic Park III*, 3D printing of bodily structures was predicted in *The Fifth Element* (Dir. Luc Besson, 1997) using a machine closely resembling a CT scanner. Tablet-style devices were depicted in Stanley Kubrick's 1968 film *2001: A Space Odyssey*, a fact used by the defence team for Samsung in 2011 when Apple sued for intellectual property theft in relation to design

FIGURE 8.19   A periscope being used above an operation which is projected onto a lantern screen for a lecture in the adjoining room. Gouache painting by W.R. Seton. Credit: Wellcome Collection. Attribution 4.0 International (CC BY 4.0).

FIGURE 8.20   A cartoon from 1932 predicting remote healthcare consultations conducted via TV screen. Credit: Line block after D.L. Ghilchip, 1932. Wellcome Collection. Attribution 4.0 International (CC BY 4.0).

FIGURE 8.21   A halftone illustration from an early radiology textbook, showing doctors huddled around their patient – viewing the interior of the chest to determine the location of a bullet using X-ray screening as depicted here is almost a direct continuation of the physical examination. 1900. Credit: Surgeons examining a mauser bullet in a man's chest via the use of an X-ray. Halftone, 1900, after W. Small. Wellcome Collection. Attribution 4.0 International (CC BY 4.0).

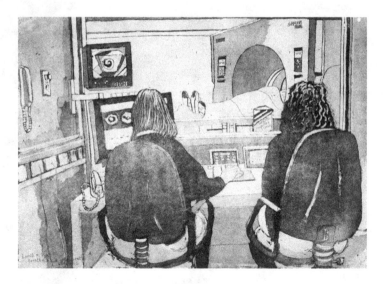

FIGURE 8.22   Etching depicting a patient entering a computerised tomography scanner, seen from the control room. Two women operating the scanner sit behind a screen and speak instructions to the patient. They have two monitors in the control room to enable them to see the scans. Credit: Etching and aquatint by Virginia Powell, ca. 1995. Wellcome Collection. Attribution 4.0 International (CC BY 4.0).

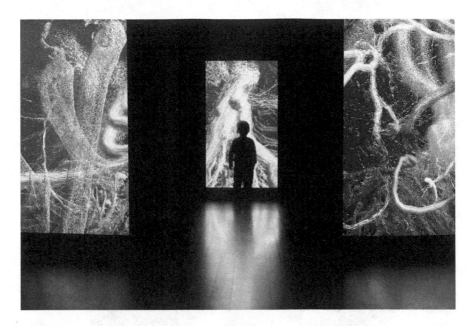

FIGURE 8.23 *The Tides Within Us* by Marshmallow Laser Feast. Image courtesy of Marshmallow Laser Feast in partnership with Fraunhofer MEVIS.

similarities between their respective tablets.[21] James Bond used a table-sized touch-screen tablet in Quantum of Solace (dir. Marc Forster, 2008) before similar devices were marketed as an interactive educational tool for displaying imaging studies. On this trajectory, I am eagerly awaiting a whizzy imaging display tool of the sort used by Tom Cruise in *Minority Report* (dir. Steven Spielberg, 2002).

"I'M NOT A ROBOT" … YET

Meanwhile, back in the real world, as part of NHS infection control measures I am required to complete a lateral flow test for the SARS-CoV-2 virus twice a week. On completing the test I then need to report the result via an online register, and as with so many online submissions these days to progress to the next page I need to click on a box saying "I'm not a robot". This declaration feels vaguely comical, a little like the lawyer who needed to clarify that he was not, in fact, a cat during online court proceedings in which his Zoom settings had turned his digital likeness into that of a wide-eyed kitten avatar.[22] Yet the fragmented, frequently pixelated interactions most of us have become accustomed to during lockdown video calls have at times made our physical reality difficult to fully separate from the digital realm. An incompletely loaded MRI brain image from an online meeting, emblematic of this digital hinterland, is shown in Figure 8.24.

Ticking the "I'm not a robot" box also provokes some unease for other reasons. For the time being, I am able to convince a computer that I am not another computer

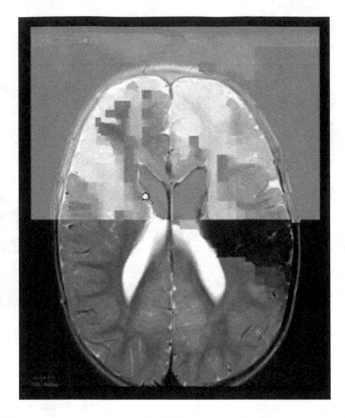

FIGURE 8.24   An incompletely loaded MRI brain image from an online meeting. Image provided by the author.

or algorithm (a sort of reversed Turing test), but it seems likely that there will come a time that the algorithms will be just as good – if not better – at spotting traffic lights or tractors in a fragmented image. When that time comes, I wonder if ticking the "not a robot" box will be an admission of being underqualified to do the job of analysing medical images.

We saw in Chapter 6 that medical images can potentially promote a mechanistic view of the body. Take a look at Figures 8.25, 8.26, 8.27, and 8.28 for some more examples of the body depicted as a robotic, mechanical device seemingly constructed from interchangeable components. We live in an era in which our digital lives can be difficult to untangle from our "real" lives, and in which "deep-fakes" may cause us to question the veracity of any screen-based image. The irony that the design of machines currently threatening people's livelihoods is based on our own neural architecture is small comfort to those most at risk, as the original human "computers" (employed to perform repetitive mathematical calculations before their machine counterparts made them redundant in the 1950s and 60s) will attest. Collectively, it is easy to conclude that the mechanistic view is correct and that we are no better than robots. The appearance of a highly erudite article in the Guardian newspaper

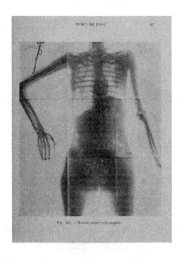

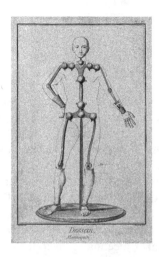

FIGURES 8.25 A composite of nine radiographs from an early radiology textbook, showing the torso and upper limbs of a human figure. Credit: Traité de radiographie médicale et scientifique: cours libre professé a l'écolepratique de la Faculté de Médecine de Paris, deuxième semestre de 1896–1897/ par Le Dr. Foveau de Courmelles; précédé d'une préface de A. d'Arsonval; avec 176 figures dans le texte. Wellcome Collection. Public Domain Mark.

FIGURE 8.26 Line drawing of a mannequin/lay figure, showing the internal structure and positioned in a similar pose to the figure in 8.25. Credit: Engraving by B.L. Prevost after Goussier. Wellcome Collection. Attribution 4.0 International (CC BY 4.0).

FIGURE 8.27 Coronal CT image of mannequin or "phantom" used as a proxy for a patient in a simulation training exercise. Image provided by the author.

FIGURE 8.28 Renaissance-era anatomical écorché figure, with right upper limb detached and moved to the side, providing an unobscured view of the torso and hip area. Credit: Engraving by Philip Galle. Wellcome Collection. Attribution 4.0 International (CC BY 4.0).

generated by an AI software algorithm has not helped me shift this idea as I struggle to construct my own coherent sentences.[23]

THE HUMAN FACTOR

There would certainly seem to be room for improvement in terms of human performance. Computers would not be required to take regular viral antigen tests, being incapable of contracting a viral infection, or at least not ones communicable to humans. Likewise, line managers would be impressed by the lack of sick days or need for annual leave, and no dip in performance just before lunch or at the end of the day. Even when fighting fit, fed, and watered, and after a good night's sleep, human radiologists remain prone to a wide range of potential sources of bias or error.

Periodic rebranding of departmental meetings primarily concerned with error in reports is reminiscent of the "euphemism treadmill" we encountered in Chapter 6. "Mistakes meeting" transformed into a "discrepancy meeting", before becoming a "learning from discrepancies meeting". In the UK the most recent incarnation is the "Radiology Events and Learning Meeting".[24] While the growing emphasis on the educational dimension of such meetings is welcome, the sensitivity related to terms such as "mistakes" and "discrepancies" is revealing.

A review article lists some ten different types of cognitive bias to which radiologists are prone,[25] including:

- *Confirmation bias*: The radiologist actively looks for features which support their hunch, whilst overlooking or ignoring features that might suggest an alternative diagnosis.
- *Satisfaction of search*: The radiologist correctly identifies an abnormality, but then does not look sufficiently thoroughly for additional abnormalities elsewhere in the scan.
- *Satisfaction of report*: An error made in a previous report is perpetuated in subsequent reports, based on the assumption that prior reports are accurate.
- *Availability bias*: The radiologist is biased by recent experience, such as overcalling a particular pathology after being made aware of missing it in a previous case, or suggesting more unusual diagnoses after reading a journal article or attending a lecture.

Radiologists are aware of these potential pitfalls and therefore often need to engage in an internal thought process (known as "metacognition") to try and factor such potential biases into their assessment. In addition to this internal battle, radiologists may from time to time be gently encouraged to massage the report wording to match the assessment of the clinician caring for the patient. In my own experience this has not been particularly problematic but there have certainly been occasions in which "clarifications" have been requested, most typically to downplay or normalise equivocal appearances. Such internal and external dialogues mean the final radiology report content is substantially more malleable than that likely to be produced by a rigid computer algorithm.

It is therefore appealing to assume that artificial intelligence, machine learning and "black box" deep neural networks are the route out of such human errors, offering unbiased and accurate information. The difficulty, of course, is that even computers that are enabled to "think for themselves" have, in some capacity, been programmed by humans – just as prone to error and bias as the rest of us. There are means of overcoming such problems, but recent studies in relation to racial bias in facial recognition software demonstrate how seemingly neutral software algorithms are all too vulnerable to human prejudices.[26] The integration of software algorithms into situations in which lives are at stake also requires extreme caution, as experience within the aviation industry demonstrates. The phrase "black box" in the context of AI now has unfortunate connotations, a reminder of the black box flight recorders retrieved from the wreckage of Lion Air flight JT610 in October 2018 (346 fatalities) and Ethiopian Airlines flight ET302 in March 2019 (157 fatalities). Both these Boeing 737 MAX utilised a piece of AI software – the "Maneuvering Characteristics Augmentation System" (MCAS) – thought to have contributed to both crashes.[27]

Within my own department the main example of machine learning in clinical use is to be found in a less situation-critical setting – a software algorithm for calculating the "bone age" of a patient, based on the appearances of a single hand radiograph (shown in Figure 8.29). This has been a welcome addition, largely replacing a rather tedious and time-consuming task of individually scoring numerous bones and toting up scores, and has definitely freed up reporting time. The software is not perfect however, sometimes misregistering images or failing to produce a plausible result, so the reports and radiographs are still checked by a radiologist both to ensure the software has worked properly and also to ensure there are not any unexpected findings that the algorithm is not programmed to detect. As yet, the algorithms have not put me out of a job.

All the same, it seems likely that computers and artificial intelligence will, in time, transform and perhaps even replace the traditional role of the radiologist. In reality, the role of the radiologist has always been in a state of ongoing evolution as technology moves forward – numerous imaging techniques have become redundant as less invasive, safer, or more accurate alternatives have replaced them. A recent paper describes the transformation of the role of the radiologist as a result of advances in genetic sequencing technology. Traditionally, clinical colleagues would request radiographs of a child with suspected skeletal dysplasia asking the radiologist to look for characteristic features in the appearance of the bones which may provide a diagnosis. Increasingly, genomic screening is providing the clinician with a diagnosis prior to radiographs being performed, such that the role of the radiologist is now shifted to confirm whether features in keeping with this genetic diagnosis are present or not, a process described as "reverse radiology".[28] The interplay between advancing genetic sequencing technology, molecular manipulation, and advancing imaging technology is eroding traditional boundaries between diagnostic examinations, disease assessment and treatment, indicated by the increasing popularity of the term "theragnostic imaging".

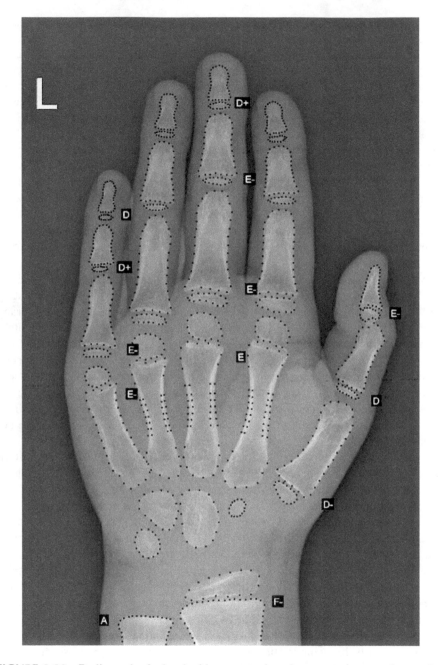

FIGURE 8.29   Radiograph of a hand with computer-based segmentation superimposed as part of an automated "bone age" calculation algorithm. Image provided by the author.

THE FALSE IMAGING HIERARCHY

Most art critics and historians would agree that the value of a work of art has little to do with how detailed, ornate or large it is – a few choice strokes of pencil or charcoal may manage to convey a great deal more than a vast elaborate oil painting. This is, of course, not the same as saying that vast oil paintings lack power or value, but the maxim that "less is sometimes more" often proves to be the case in the artistic realm. In contrast to the art world, in medicine there is often a perception of a hierarchy of imaging, with less expensive and widely available imaging technology such as plain radiography and ultrasound considered basic (perhaps akin to a speedy charcoal sketch), conventional CT and MRI next up the chain (a well-rendered watercolour?) and highly specialised scanners of limited availability such as PET-CT or PET-MRI at the top of the ladder (a massive oil painting in a fancy gold frame). The hierarchy follows the logical sequence of how clinical medicine has traditionally been delivered – taking a clinical history from the patient first, then performing physical examination, then simple "bedside" tests such as urine dipsticks, then basic laboratory tests such as common blood tests and so forth. In the UK this model also reflects the hierarchical structure of healthcare provision that a patient might need to follow for treatment of a complicated disease, visiting their general practitioner first before being referred for further evaluation in a district general hospital, then onto a tertiary referral centre and on occasion a quaternary centre. It is important to recognise that the skills and knowledge of medical staff at each step are not hierarchical, with expertise tailored to the specific role undertaken. Likewise, particular imaging investigations cannot be readily labelled basic to advanced in sequential order – each investigation has a particular profile with different advantages, disadvantages, and safety aspects to consider. The supposed imaging hierarchy certainly bears little resemblance to the benefit a patient may receive from any given imaging investigation. A law of diminishing returns often applies; a "simple" (certainly low cost and widely available) chest radiograph can provide life-saving information, whilst many individuals will gain little from the most sophisticated multi-modality scanners (high cost, very high carbon footprint). The difficulty of this apparent hierarchy is the perception that rather than settling with a "basic" test (such as ultrasound), a higher level of certainty and accuracy will be achieved with a more "advanced" test (perhaps CT or MRI). While this may often be the case, the pervasive idea that some imaging investigations are "better" than others rather than "different", risks exposing the patient to unnecessary imaging.

THE IMAGING INDUSTRIAL COMPLEX?

A Google search for the phrase "Pharmaceutical industrial complex" returns 115,000,000 hits, but as of March 2021 I have not found any hits for the phrase "imaging industrial complex", so I think I might stake a claim for coining this phrase. Eisenhower warned of the "military industrial complex" in his farewell address to the United States in 1961[29] and in recent decades there has been growing concern in relation to the "prison industrial complex".[30] I would certainly not claim medical imaging is in the same league as these notorious examples, but any undertaking delivered on an industrial scale tends to promote self-sustaining patterns of activity. Certainly, in healthcare systems in which a radiologist is paid on a fee-for-examination basis, there is a potential incentive for recommending further imaging studies regardless of whether they are really required. Working within the NHS, there is no direct benefit to me as a radiologist in recommending additional examinations, and evidence-based guidelines such as those produced by the Royal College of Radiologists help to ensure imaging is used appropriately according to clinical need.[31] Nevertheless, the number of examinations reported by a radiologist per year will still be scrutinised by line managers as a marker of productivity, and at a departmental level the number of examinations undertaken has implications for future funding. At a strategic level the concept of doing less imaging becomes unthinkable. Accordingly, radiology departments in nearly all healthcare systems become highly efficient at performing imaging examinations, but increasingly less efficient at identifying and eliminating inappropriate requests for imaging. There is also an element of "mission-creep" as scans which were originally used as a problem-solving tool in difficult clinical cases, become used more widely, ultimately becoming a screening tool. This is well illustrated by the use of CT, which is now increasingly used in the diagnosis of acute appendicitis in adults, having previously been diagnosed on the basis of the clinical history and examination. The trend becomes self-perpetuating as the criteria for a scan gradually shifts from high probability of appendicitis (although not high enough to make the diagnosis on clinical grounds alone), to moderate probability, to possibility, to every patient with abdominal pain. For imaging departments based on a fee-for-service model, more scans mean more money so a lowering of the threshold by which clinicians use to request scans – intentional or otherwise – is potentially lucrative.

When there is a tendency to think technological solutions are the answers to all our problems it is instructive to consider an episode of medical history that occurred more than 50 years prior to the discovery of X-rays. In 1847 eminent German physician Rudolf Virchow was sent to investigate the causes of a typhus epidemic that had broken out in Upper Silesia. The Prussian authorities had anticipated a medical solution, but Virchow's report advocated "political medicine" urging education, freedom and prosperity, particularly for the oppressed Polish minority, who were disproportionately affected by the epidemic.[32] Virchow's findings have striking contemporary resonance in the light of health inequalities exposed by the coronavirus pandemic, reminding us of the importance of the broader determinants of health.

IMAGING WITHOUT IMAGES?

In the selection of illustrations in this chapter and throughout the book I have emphasised the millennia-old tradition of looking within the body's interior as a cornerstone of medical science as well as an enduring source of fascination and spectacle. Yet in the same way that coins may soon cease to jangle in people's pockets and purses, the days of medical images constructed specifically for human eyes may also be numbered.

We are taking steps towards this era in the form of deep neural networks in which vast quantities of imaging data are poured into the computer at one end, and (in theory) useful information which can be acted upon comes out at the other end. The processes occurring between these steps are becoming increasingly difficult to make sense of, leading to calls for such algorithms to be more transparent and understandable (a shift from "black box" to "glass box" processes). Nevertheless, it may be that if the answers the computers come up with are sufficiently reliable, there may be a loss of interest in how it is done. Cutting out the unnecessary intermediary step of constructing images tailored to human vision, the scanners of the future may well produce answers without pictures. Although counterintuitive this does take medical imaging to a logical conclusion. Whereas art is fundamentally about conveying meaning, the purpose of medical imaging is to extract meaning within the very specific framework of helping to improve an individual's health.

In this scenario – admittedly a little way off for now – healthcare providers will need to think carefully about what questions to ask the scanner, much like consulting the oracle in classical mythology. Perhaps by this point scanning technology is sufficiently safe that screening patients before they become symptomatic will be feasible not for a few select conditions but for all diseases and that it can be done on a population level. There is a scene in the Disney CGI movie *Big Hero Six* (dir. Don Hall and Chris Williams, 2014) in which Baymax, an advanced healthcare robot, scans the entire population of the city of San Fransokyo in a matter of seconds, acquiring not only structural anatomical information of the kind I am familiar with looking at but also biochemistry including hormonal levels etc. What to do with all this information though?

In current radiological practice imaging is provided to help guide the management of individual patients – if the examination won't change (or potentially change) the management of the patient, we shouldn't be doing the scan. But if instead of scanning individual patients with specific management questions to answer, we are scanning on a city-wide or perhaps population-wide level, what questions should we be seeking to answer? The ones I imagine ought to be high on the list are: what is required for this population to lead healthier, longer lives? What will improve quality of life?

Yet of course we don't need to wait for science fiction to become science reality in this regard, we already know the answers:

- Ensure air is safe to breathe (4.2 million deaths every year occur as a result of exposure to outdoor air pollution)[33]
- Ensure water is safe to drink (3.6 million deaths every year from water-related diseases)[34]
- Improve nutrition (each year 3.1 million children die from undernutrition,[35] 2.8 million people die as a result of being overweight or obese)[36]
- Reduce (or eliminate) tobacco consumption (7 million deaths per year from direct tobacco use, 1.2 million deaths per year from second-hand smoke)[37]
- Reduce (or eliminate) alcohol consumption (3 million deaths every year result from harmful use of alcohol)[38]

- Improve road safety (1.35 million people die each year as a result of road traffic crashes[39] – mandatory speed limiters in all motor vehicles would be my suggestion)
- Facilitate regular physical exercise (3.2 million deaths per year from physical inactivity)[40]

Sadly, of course, this list could go on (gun control, addressing male aggression and sexual violence, building housing out of non-flammable materials) and cumulatively these issues may appear insurmountable. The limitations of what medical imaging can currently achieve in the face of such global health challenges were recently impressed upon me by two graphs – one showing a near exponential rise in the use of cross-sectional imaging in Scotland between 2009–2015,[41] the other showing life expectancy plateauing during the same period, and indeed beginning to dip between 2015 and 2019.[42]

Rudolf Virchow reminds us that the solutions to improving health are often political, not technological. The causative link between tobacco and lung cancer was well established by the 1950s.[43] I find it bewildering that some 70 years later CT scanning is being utilised to screen for early stage lung cancers, the large majority of which are caused by smoking (directly or otherwise). Evidence is increasingly persuasive that such programmes could save lives,[44, 45] but in the context of finite healthcare resources, it is imperative that we examine the broader trend of using such resource-hungry downstream solutions to address preventable health problems. Going forwards, technology-heavy approaches to dealing with diseases that we already know how to prevent are simply not sustainable.

The resources that modern medical imaging consumes are not confined to healthcare budgets. The carbon footprint and environmental impact of medical imaging technology are substantial, and is disproportionately skewed by "high end" scanners such as CT, MRI, and PET.[46] It is also important to recognise that the lion's share of the workload performed using such energy-hungry machines relates to oncology and cardiovascular disease, much of which is potentially preventable. Data suggesting one in five deaths in 2018 were caused by burning fossil fuels[47] should compel any industry to address its energy consumption, particularly the healthcare sector in which the ethical principles of justice and non-maleficence should be paramount.

While the origins of the SARS-CoV-2 virus are still being investigated at the time of writing, there is a body of opinion suggesting environmental degradation is likely to have been contributory to the emergence of this virus[48] which has claimed the lives of over 4.1 million people worldwide at the time of writing.[49] As of 2019, the HIV pandemic has resulted in 32.7 million deaths from AIDS-related illnesses.[50] The jump of HIV from chimpanzees to humans has likewise been linked to deforestation and interference in fragile ecosystems.[51]

IMAGINING SOLUTIONS

This all seems like a thoroughly depressing way in which to finish off this book. X-rays are often known as "the invisible light", so I hope I can convince you there may be a chink of light at the end of the tunnel, even if we are unable to see it. Knowledge, most agree, is power and humanity currently enjoys an extraordinary wealth of scientific information that empowers us to address the vast global challenges confronting us. My thought experiment demonstrates that we do not need to wait for AI scanners to tell us how to improve health, either at individual or population level. In the UK, large-scale use of financial resources and restrictions to personal freedoms enacted by a Conservative government to save lives during the COVID-19 pandemic puts pay to "nanny state" arguments against public health interventions. If a life is worth saving from a viral aetiology, surely the same applies to any established hazard to human health.

Taking a more positive approach to the statistics can also be instructive. A recent paper analysing data from 168 countries demonstrated that regular physical exercise prevents 3.9 million deaths per year,[52] so comparing that to the 3.2 million deaths attributed to physical inactivity, the proverbial glass may be slightly fuller than empty. The paper also provides a striking message – active participation is required to save lives.

These concluding thoughts could be characterised as looking at the "big picture", a phrase reminding us once more of the metaphorical power of images. In all their forms images have the power to inspire, enrage, arouse, and disgust. Medical images, despite some of the associated difficulties I have highlighted, do genuinely save many, many lives. Seen in this light, they can claim to be among the most powerful images ever produced by humanity. Yet their power also resides in belonging within the broader tradition of representing the body. Images generated on an industrial scale may sit uneasily labelled as art, and in serving a specific purpose medical images certainly do not meet Oscar Wilde's understanding of the term.[53] Nevertheless, such images are undeniably shaped by, and interpreted using, techniques and templates originating within the artistic realm. Cross-pollination between different disciplines has been an integral and inseparable part of the development of medical imaging technology, a theme repeated in many endeavours in the history of science.

The development of several successful COVID-19 vaccines in a shorter period of time than it has taken me to write this book is a testament to the power of modern medical science, and would seem to owe little to the creative industries. Convincing the public to take the vaccine, however, requires all manner of creative endeavour not typically considered within the umbrella of medical science. Indeed, medicine has long been characterised as an art as well as a science, and medical imaging certainly has a foot planted firmly in both camps. Tackling immense global challenges such as poverty, inequality, and the climate crisis will likewise require interdisciplinary cooperation, partnership and teamwork on an unprecedented scale. Most of all it will require imagination.

REFERENCES

1. S. Avery, "A new kind of rays": Gothic fears, cultural anxieties and the discovery of X-rays in the 1890s, *Gothic Stud.* 17(1), May 2015, 61–75.
2. L. Cartwright, *Screening the Body: Tracing Medicine's Visual Culture*, University of Minnesota, Minnesota, 1995, p 114.

3. A. M. K. Thomas and A. K. Banerjee, *The History of Radiology*, Oxford University Press, Oxford, 2013, p 69.

4. X-Rays, directed by George Albert-Smith, 1897, BFI Player, https://player.bfi.org.uk/free/film/watch-x-rays-1897-online, last accessed 20/07/21.

5. Wikipedia, A computer animated hand, https://en.wikipedia.org/wiki/A_Computer_Animated_Hand, last accessed 20/07/21.

6. Twitter message posted by PIXAR (@PIXAR, the official Twitter for PIXAR), 19/05/17, https://twitter.com/i/status/865357349027852288, last accessed 20/07/21.

7. P. S. Calhoun, B. S. Kuszyk, D. G. Heath, J. C. Carley and E. K. Fishman, Three-dimensional volume rendering of spiral CT data: Theory and method, *RadioGraphics* 19(3), 1999, 745–764.

8. Lisa Zyga, Quantum no-hiding theorem experimentally confirmed for first time, *Phys.org*, https://phys.org/news/2011-03-quantum-no-hiding-theorem-experimentally.html#, last accessed 20/07/21.

9. Harald Sack, Giuseppe Fiorelli's excavations in Pompeii, *SciHi.org*, 08/06/17, http://scihi.org/giuseppe-fiorelli/, last accessed 20/07/21.

10. J. A. B. Collier (ed), *Oxford Handbook of Clinical Specialties, 5th edition*. Oxford University Press, Oxford, 1999.

11. Robert Krulwich, Cosmonaut crashed into earth 'Crying In Rage', *NPR*, 18/03/11, https://www.npr.org/sections/krulwich/2011/05/02/134597833/cosmonaut-crashed-into-earth-crying-in-rage, last accessed 20/07/21.

12. Hannah Rose Mendoza, 3D printing (re)produces 65 million year old mating call, *3D Print*, 19/11/15, https://3dprint.com/106320/3d-printed-dino-skull-sound/, last accessed 20/07/21.

13. D. M. Howard, J. Schofield, J. Fletcher et al. Synthesis of a vocal sound from the 3,000 year old mummy, Nesyamun 'True of Voice', *Sci Rep* 10, 2020, 45000. doi:10.1038/s41598-019-56316-y

14. BBC News, Mummy returns: Voice of 3,000-year-old Egyptian priest brought to life, *BBC News*, 24/01/21, https://www.bbc.co.uk/news/world-middle-east-51223828, last accessed 20/07/21.

15. R. Richardson, *Death, Dissection and the Destitute*. 2nd edition, University of Chicago Press, Chicago, 2001.

16. L. Cartwright, *Screening the Body: Tracing Medicine's Visual Culture*, University of Minnesota, Minnesota, 1995, p 115.

17. Neil Versel, Ohio state surgeon streams operation on periscope for medical education, *MedCity News*, 27/06/15, https://medcitynews.com/2015/06/ohio-state-surgeon-streams-operation-on-periscope-for-medical-education/, last accessed 20/07/21.

18. Denis Campbell, GPs prefer to see patients face to face, says UK family doctors' leader, *Guardian*, 28/03/21, https://www.theguardian.com/society/2021/mar/28/gps-prefer-to-see-patients-face-to-face-says-uk-family-doctors-leader, last accessed 20/07/21.

19. Guardian, Conjoined twin girls separated in London after 50 hours of surgery, *Guardian*, 15/07/19, https://www.theguardian.com/society/2019/jul/15/conjoined-twins-separated-great-ormond-street-safa-marwa#, last accessed 20/07/21.

20. Mark Westall, The tides within us, *FAD Magazine*, 14/12/20, https://fadmagazine.com/2020/12/14/the-tides-within-us/, last accessed 20/07/21.

21. Ned Potter, Stanley Kubrick envisioned the iPad in '2001,' says Samsung, *ABC News*, 26/08/11, https://abcnews.go.com/Technology/apple-ipad-samsung-galaxy-stanley-kubrick-showed-tablet/story?id=14387499, last accessed 20/07/21.

22. Independent, Lawyer assures judge he isn't a cat after Zoom filter faux pas, *Independent*, https://www.independent.co.uk/tv/news/lawyer-assures-judge-he-isn-t-a-cat-after-zoom-filter-faux-pas-vd7a51f69, last accessed 20/07/21.

23. Guardian, GPT-3, A robot wrote this entire article. Are you scared yet, human? *Guardian*, 08/09/20, https://www.theguardian.com/commentisfree/2020/sep/08/robot-wrote-this-article-gpt-3, last accessed 20/07/21.
24. G. Maskell, Virtual special issue: Error in radiology, *Clin Radiol.* 75, 2020, 159–160.
25. L. P. Busby, J. L. Courtier and C. M. Glastonbury, Bias in radiology: The how and why of misses and misinterpretations, *RadioGraphics*, 38, 2018, 236–247. doi:10.1148/rg.2018170107
26. J. Lunter, Beating the bias in facial recognition technology, *Biometric Technol. Today* 2020(9), 2020, 5–7.
27. Dominic Gates, Q&A: What led to Boeing's 737 MAX crisis, *Seattle Times*, 18/11/20, https://www.seattletimes.com/business/boeing-aerospace/what-led-to-boeings-737-max-crisis-a-qa/, last accessed 20/07/21.
28. A. C. Offiah and C. M. Hall, The radiologic diagnosis of skeletal dysplasias: Past, present and future. *Pediatr Radiol* 50, 2020, 1650–1657.
29. Wikipedia, Eisenhower's farewell address, https://en.wikipedia.org/wiki/Eisenhower%27s_farewell_address, last accessed 20/07/21.
30. Wikipedia, Prison–industrial complex, https://en.wikipedia.org/wiki/Prison%E2%80%93industrial_complex, last accessed 20/07/21.
31. iRefer, Making the best use of clinical radiology, 8th edition, Royal College of Radiologists, online resource, https://www.irefer.org.uk/, last accessed 20/07/21.
32. R. Porter, *The Greatest Benefit to Mankind: A Medical History of Humanity from Antiquity to the Present*, HarperCollins, London, 1997, p 415.
33. World Health Organisation, Air pollution, https://www.who.int/health-topics/air-pollution#tab=tab_1, last accessed 20/07/21.
34. World Health Organisation, Drinking-water, 14/06/19, https://www.who.int/news-room/fact-sheets/detail/drinking-water, last accessed 20/07/21.
35. World Hunger, World child hunger facts, https://www.worldhunger.org/world-child-hunger-facts/, last accessed 20/07/21.
36. World Health Organisation, Obesity, 09/06/21, https://www.who.int/news-room/facts-in-pictures/detail/6-facts-on-obesity, last accessed 20/07/21.
37. World Health Organisation, Tobacco, 27/05/21, https://www.who.int/news-room/fact-sheets/detail/tobacco, last accessed 20/07/21.
38. World Health Organisation, Alcohol, 21/09/18, https://www.who.int/news-room/fact-sheets/detail/alcohol, Last accessed 20/07/21.
39. World Health Organisation, Road traffic injuries, 21/06/21, https://www.who.int/news-room/fact-sheets/detail/road-traffic-injuries, last accessed 20/07/21.
40. NCD Alliance, Physical activity, https://ncdalliance.org/why-ncds/ncd-prevention/physical-activity, last accessed 20/07/21.
41. Royal College of Radiologists, Clinical radiology UK workforce census 2015 report, https://www.rcr.ac.uk/publication/clinical-radiology-workforce-scotland-2015-census-report (p 15), last accessed 20/07/21.
42. Healthy life expectancy in Scotland, 2017-2019, *National Records of Scotland report*, https://www.nrscotland.gov.uk/statistics-and-data/statistics/statistics-by-theme/life-expectancy/healthy-life-expectancy-in-scotland/2017-2019, last accessed 20/07/21.
43. Robert N. Proctor, The history of the discovery of the cigarette-lung cancer link: Evidentiary traditions, corporate denial, global toll, *Tobacco Control,* 21, 2012, 87e91. doi:10.1136/tobaccocontrol-2011-050338
44. National Lung Screening Trial Research Team, D. R. Aberle, A. M. Adams, et al., Reduced lung-cancer mortality with low-dose computed tomographic screening. *N Eng J Med.* 365, 2011, 395–409.

45. Denis Campbell, CT scan catches 70% of lung cancers at early stage, NHS study finds, *Guardian,* 14/02/21, https://www.theguardian.com/society/2021/feb/14/ct-scan-catc hes-70-of-lung-cancers-at-early-stage-nhs-study-finds, last accessed 20/07/21.

46. E. Picano, Environmental sustainability of medical imaging, *Acta Cardiologica*, 9, 2020, 1–5. doi:10.1080/00015385.2020.1815985, Online ahead of print. PMID: 329 01579.

47. K. Vohra, A. Vodonos, J. Schwartz, E. A. Marais, M. P. Sulprizio and L. J. Mickley, Global mortality from outdoor fine particle pollution generated by fossil fuel combustion: Results from GEOS-Chem, *Environ. Res.* 195, 2021, 110754, ISSN 0013-9351.

48. Aaron Bernstein, Coronavirus and climate change, Harvard T.H. Chan, School of Public Health, https://www.hsph.harvard.edu/c-change/subtopics/coronavirus-and-c limate-change/, last accessed 20/07/21.

49. Coronavirus death toll (4,115,872 deaths), Worldometer, https://www.worldometers. info/coronavirus/coronavirus-death-toll/, last accessed 20/07/21.

50. Global HIV & AIDS statistics — Fact sheet, UN AIDS, https://www.unaids.org/en/ resources/fact-sheet, Last accessed 20/07/21.

51. J. Pépin, *The Origins of AIDS*. 2nd edition, Cambridge University Press, Cambridge, 2021.

52. T. Strain, S. Brage, S. J. Sharp, J. Richards, M. Tainio, D. Ding, J. Benichou and P. Kelly, Use of the prevented fraction for the population to determine deaths averted by existing prevalence of physical activity: a descriptive study, *Lancet Glob Health*, 8, 2020, e920–30

53. O. Wilde, "Art is useless because..." 1891, Letter No. 123, p344, in *Letters of Note: Correspondence Deserving of a Wider Audience*, Compiled by S. Usher, Canongate Books, London, 2013.

Index

Note: Locators in *italics* represent figures in the text.

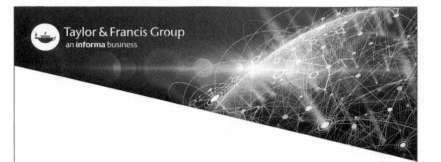

CPSIA information can be obtained
at www.ICGtesting.com
Printed in the USA
BVHW022150260223
659179BV00016B/277